ENVIRONMENTAL HEALTH - PHYSICAL, CHEMICAL AND BIOLOGICAL FACTORS

HARMFUL ALGAL BLOOMS AND HYPOXIA

U.S. ASSESSMENTS, RESEARCH PLAN AND ACTION STRATEGY

Environmental Health - Physical, Chemical and Biological Factors

Additional books in this series can be found on Nova's website under the Series tab.

Additional e-books in this series can be found on Nova's website under the eBooks tab.

ENVIRONMENTAL HEALTH - PHYSICAL, CHEMICAL AND BIOLOGICAL FACTORS

HARMFUL ALGAL BLOOMS AND HYPOXIA

U.S. ASSESSMENTS, RESEARCH PLAN AND ACTION STRATEGY

ROBERTA ALVAREZ
EDITOR

nova publishers
New York

Library of Congress Cataloging-in-Publication Data

ISBN: 978-1-63485-399-6

Published by Nova Science Publishers, Inc. † New York

CONTENTS

PREFACE

Harmful algal blooms (HABs) and hypoxia (low-oxygen conditions) in the marine waters and freshwaters of the United States negatively impact resources across thousands of miles of the Nation's coastal and inland waters. Their prevalence and duration encompass scientifically complex and economically damaging aquatic issues. This book assesses threats posed by HABs and hypoxia and provides a comprehensive research plan and action strategy outlining Federal agencies' roles and responsibilities for evaluating and managing them.

In: Harmful Algal Blooms and Hypoxia
Editor: Roberta Alvarez

ISBN: 978-1-63485-399-6

Chapter 1

HARMFUL ALGAL BLOOMS AND HYPOXIA COMPREHENSIVE RESEARCH PLAN AND ACTION STRATEGY: AN INTERAGENCY REPORT*

Interagency Working Group on the Harmful Algal Bloom and Hypoxia Research and Control Act

EXECUTIVE OFFICE OF THE PRESIDENT
NATIONAL SCIENCE AND TECHNOLOGY COUNCIL
WASHINGTON, D.C. 20502

February 11, 2016

Dear Colleagues:

I am pleased to transmit to you *Harmful Algal Blooms and Hypoxia Comprehensive Research Plan and Action Strategy: An Interagency Report*, a report to Congress produced by the Interagency Working Group on the Harmful Algal Bloom and Hypoxia Research and Control Act (IWG-HABHRCA). IWG-HABHRCA is organized under the National Science and

* This is an edited, reformatted and augmented version of a report issued by The Office of Science and Technology Policy, February 2016.

Technology Council; Committee on Environment, Natural Resources, and Sustainability; Subcommittee on Ocean Science and Technology.

The 2014 reauthorization of the Harmful Algal Bloom and Hypoxia Research and Control Act (HABHRCA 2014; P.L. 113-124) acknowledges continued concerns related to harmful algal blooms (HABs) and hypoxia, extends the scope of the legislation to include freshwater HABs and hypoxia, and recognizes the need for further coordinated Federal action to address these issues. Specifically, the legislation calls for a “comprehensive research plan and action strategy” and assessments of harmful algal blooms and hypoxia, as well as an “integrated assessment on Great Lakes hypoxia and harmful algal blooms.” This report fulfills these requirements.

Sincerely,
John P. Holdren
Assistant to the President for Science and Technology Director,
Office of Science and Technology Policy

EXECUTIVE SUMMARY

The prevalence and duration of harmful algal blooms (HABs) and hypoxia (low-oxygen conditions) in the marine waters and freshwaters of the United States, including the Great Lakes, are generating public concern. From extended shellfish closures on the west coast in 2015, to a larger-than-predicted hypoxic zone in the Gulf of Mexico, these events negatively impact resources across thousands of miles of the Nation’s coastal and inland waters, and represent some of the most scientifically complex and economically damaging aquatic issues. HABs and hypoxia pose a significant challenge to the ability to safeguard the health of the Nation’s coastal and freshwater ecosystems.

HABs and hypoxia have serious effects on a community’s social and public health. They may threaten the safety of seafood, drinking water, and air quality. HABs and hypoxia events may also result in lost revenue for lakefront and coastal economies that are dependent on aquatic/seafood harvest or tourism, disruption of subsistence activities, loss of community identity tied to aquatic-resource use, and disruption of social and cultural practices.

Harmful Algal Bloom and Hypoxia Research and Control Act

In June 2014, Congress reauthorized the Harmful Algal Bloom and Hypoxia Research and Control Act (HABHRCA) by passing the Harmful Algal Bloom and Hypoxia Research and Control Amendments Act of 2014 (HABHRCA 2014, P.L. 113-124). The reauthorization of HABHRCA acknowledged concerns related to HABs and hypoxia, extended the scope of the legislation to include freshwater HABs and hypoxia, and recognized the need for further coordinated action across the Federal sector to address these issues. Additionally, the legislation called for Federal agencies to provide integrated assessments on the causes and consequences of and approaches to reducing HABs and hypoxia nationally, with particular emphasis on the Great Lakes. Finally, the reauthorization included a specific focus on the needs of stakeholders, requiring that Federal agencies engage with stakeholders around the country.

HABs in Lake Erie (Photo courtesy of the National Oceanic and Atmospheric Administration).

Federal Accomplishments

Since the original authorization of HABHRCA in 1998, and in many cases as a result of the legislation, the agencies involved in writing this report have made significant progress in addressing the causes of HABs and hypoxia, and in helping to mitigate the impacts of these issues. These achievements include:

- Establishing health advisories for freshwater toxins, specifically the cyanobacterial toxins microcystins and cylindrospermopsin;
- Developing HAB and hypoxia forecast products through more comprehensive monitoring integrated with satellite coverage and modeling of ocean, coastal, and freshwater zones, to provide time for mitigation and response;
- Developing and deploying lower cost, easy to use, and real-time sensors for early detection of hypoxia and HAB cells and toxins; and
- Improving understanding of the effects of HAB toxins on human and animal health.

Report Purpose

In recognition of the challenges presented by HABs and hypoxia, HABHRCA emphasizes the mandate to advance the scientific understanding and ability to detect, predict, control, mitigate, and respond to these types of events. This legislation establishes the Interagency Working Group on HABHRCA (IWG-HABHRCA). The legislation tasks the group with coordinating and convening Federal agencies to discuss HAB and hypoxia events in the United States, and to develop action plans, reports, and assessments of these situations.

Federal Response to HABs and Hypoxia

Federal response to HABs and hypoxia events requires considerable interagency collaboration. Some recent examples where agencies have worked together to address HABs and hypoxia are laid out below:

- *The Gulf of Mexico annually experiences a hypoxic zone that varies in size. The size of the zone informs Federal, state, and local planning and management actions in the Mississippi River basin to reduce nutrient inputs that contribute to hypoxia. The Environmental Protection Agency (EPA) and the National Oceanic and Atmospheric Administration (NOAA) collaborated to map out the 2015 "dead zone," determining that this year's zone was 6,474 square miles. It was larger than the 5,052 square miles measured last year.*

- *Since late spring 2015, the west coast of the United States has experienced an ongoing bloom of* Pseudo-nitzschia, *an algal species that can produce a toxin that impacts human and animal health. Agencies including EPA, NOAA, and the Food and Drug Administration (FDA) have worked together to monitor and analyze the bloom, and to provide event-response assistance to local and regional communities.*
- *Many agencies collaborate to forecast and establish early warning systems for HABs in Lake Erie. In 2015, the United States Geological Survey (USGS), the National Aeronautics and Space Administration (NASA), EPA, and NOAA worked together to track the development of a bloom in Lake Erie using satellites as well as collected information on water quality, biomass size, and bloom toxicity levels. These agencies established an ongoing multi-agency research effort designed provide early warnings for freshwater nuisance and toxic algal blooms by using satellites that can gather color data from freshwater bodies during scans of the Earth. The project will improve the understanding of the environmental causes and health effects of blooms driven by cyanobacteria and phytoplankton.*

This comprehensive HAB and hypoxia research plan and action strategy outlines Federal agencies' roles and responsibilities for evaluating and managing HABs and hypoxia, agency successes since the 2008 authorization of HABHRCA, and their management and response actions. This report also identifies remaining challenges and makes recommendations for actions to address these events. It draws from direct contributions from Federal agencies involved with managing and researching HABs and hypoxia, as well as from stakeholders from Federal, state, and local governments; academia; industry; nongovernmental organizations; and interested citizens. It incorporates those contributions to ensure that the recommendations align closely with and deliver products that address stakeholder needs.

Recommendations

This report recommends the following action plan for addressing HABs and hypoxia:

- Add to and improve scientific understanding of HABs and hypoxia, and their causes and effects, as well as improve testing and research methods.

Currently, some of the biggest challenges to analyzing HAB and hypoxia dynamics include the inability to predict the onset of toxicity, a need to more specifically explain the influence that excess nutrients and other factors play in the occurrence and distribution of HABs and hypoxia, a lack of studies related to toxins in foods, the need to establish methods for suppressing or controlling HABs, the need for more science to understand the influence of a changing climate on HABs and hypoxia, and the need for tools to manage a limited availability of certified reference materials (CRMs) and validated detection and analysis methods. The IWG-HABHRCA recommends the development of an interagency group specifically tasked with examining the possibility of developing CRMs and other standardized and validated detection and analysis methods.

- Strengthen and integrate new and existing monitoring programs.

  A thorough and scientifically-based monitoring program is critical to determining the location and extent of HABs and hypoxia occurrences, so that their causes can be controlled. Moreover, comprehensive monitoring is essential to providing baseline data for determining correlations of cause and effect, and is done by maintaining long-term records of environmental conditions to monitor future changes and provide data on how a changing climate may affect HAB and hypoxia events. In order to mitigate the impacts of HABs and hypoxia, it is also important to incorporate new technologies into monitoring systems that are used broadly; use observations to validate existing models; and identify key indicators that can be used to predict and identify HAB and hypoxia events at the local, regional, and national levels.

- Improve predictive capabilities by developing and enhancing HAB and hypoxia modeling programs; improve disease surveillance for human and animal exposure, illnesses, and death.

  Predictive models are critical for understanding HAB and hypoxia effects on ecosystems, and for addressing the prediction of, and response to, toxins in drinking and recreational waters. Improving and expanding modeling efforts and remote satellite sensing can help to promote technologies that minimize human and animal exposure and protect public health. Additionally, these improvements can help assess progress of management actions that inform mitigation strategies. A rapid response protocol is necessary for detecting and mitigating HAB toxins in humans and animals.

- Improve stakeholder communications, including having more effective and readily-available public advisories, stronger connections with susceptible communities, and a better understanding of the socioeconomic and health-related impacts of HABs and hypoxia.

  The first step to reducing HAB impacts on humans is to understand their causes, health effects, and how best to communicate risks associated with HAB exposure. Communication among wildlife, veterinary, medical, and public health officials, as well as with the general public, needs to improve. Additionally, more should be understood and shared amongst these groups about the socioeconomic impacts of HABs and hypoxia to local, regional, and national areas.

- Continue and expand collaborations in research, management, and policy-related arenas.

  Many of the research initiatives mentioned in this report have been made possible by collaborations between Federal agencies, as well as between these agencies and state and local entities, the public, and academia. Collaboration and information-sharing among all stakeholders vastly increases the probability that people become educated about changes in their environment and can establish early measures to mitigate the effects of HABs and hypoxia.

Many, if not all, of these recommendations are interrelated. Improved scientific understanding of HABs and hypoxia can be gained in part through improved monitoring efforts. Modeling programs are dependent upon an understanding of the dynamics of these events, as well as long-term and robust monitoring efforts. Continued collaborations are necessary to achieve any of the aforementioned goals.

The Nation's marine waters and freshwaters provide many goods and services to communities and the country. They play critical roles in the nation's transportation, economy, and security. This report outlines needs, gaps, and successes, and provides a series of recommendations for addressing HABs and hypoxia in the United States to ensure continuity of goods and services for the future.

1. Report Motivation and Purpose

The marine waters and freshwaters of the United States, including the Great Lakes, are increasingly impacted by HABs and hypoxia. These phenomena, which routinely have negative impacts on resources across hundreds of miles of American coastal and inland waters, are scientifically complex and economically damaging aquatic issues. HABs and hypoxia challenge the ability to safeguard the health of the Nation's coastal and freshwater ecosystems.

1.1. Research and Legislative Background

In the early 1980s, concern about low levels of dissolved oxygen, which aquatic species absorb and require to live (Boyd, 2015), in coastal bodies of water in the United States led to the first national assessment of coastal hypoxia (Whitledge, 1985). This assessment found declining dissolved oxygen levels in many bodies of water, in part as a result of high nutrient levels (eutrophication). By the 1990s, serious and large-scale water-quality problems, including HABs and hypoxia, had been identified, most prominently in the northern Gulf of Mexico, Lake Erie, the Chesapeake Bay, and Long Island Sound. These problems led to a national Federal assessment of eutrophication in 1999 (Bricker et al., 1999), which was updated in 2007 (Bricker et al., 2007). These problems also motivated a Federal assessment of hypoxia in the northern Gulf of Mexico (Committee on Environment and Natural Resources (CENR), 2000; EPA, 2007), and passage of the Harmful Algal Bloom and Hypoxia Research and Control Act of 1998 (HABHRCA, P.L. 105-383). The HABHRCA specifically included HABs after numerous occurrences around the country made it clear that the government needed to take action to address this increasingly prevalent phenomenon.

The HABHRCA was reauthorized as the Harmful Algal Bloom and Hypoxia Research and Control Amendments Act of 2004 (HABHRCA 2004, P.L. 108-456), which required reports that specifically addressed freshwater and marine HAB and hypoxia management to be submitted to Congress. In 2008, the Interagency Working Group on HABs, Hypoxia, and Human Health (IWG-HHHH)—tasked with responding to these requirements—submitted to Congress a HAB-management plan, "The Harmful Algal Bloom Management and Response: Assessment and Plan" (Jewett et al., 2008), which was soon followed by a more-detailed plan developed with wide community input, "The National Scientific Development, Demonstration, and Technology Transfer Plan on Reducing Impacts from Harmful Algal Blooms" (Dortch et al., 2008). The HABHRCA was reauthorized again in 2014 through the Harmful Algal Bloom and Hypoxia Research and Control Amendments Act of 2014 (HABHRCA 2014, P.L. 113-124). This reauthorization reconstituted the Interagency Task Force on HABs and Hypoxia as the Interagency Working Group on HABHRCA (IWG-HABHRCA, Table 1), which is organized under the National Science and Technology Council; Committee on Environment, Natural Resources, and Sustainability; Subcommittee on Ocean Science and Technology. The IWG-HABHRCA coordinates and convenes relevant Federal agencies to discuss HABs and hypoxia events in the United States, and develops needed reports and assessments on these topics. These assessments include updates to the two aforementioned reports produced under the last authorization. These updates have been incorporated into this report.

1.2. Purpose of This Report

This report provides Congress and stakeholders with a research plan and action strategy on HABs and hypoxia. It outlines the fundamental components of HAB and hypoxia management and response, and identifies challenges related to HABs and hypoxia events, ongoing and planned scientific research, and Federal agencies' roles and responsibilities in evaluating and managing HABs and hypoxia. The report also describes recent scientific research that is the basis for improving how the country prevents and addresses these issues.

Information contained in this report was synthesized from a number of sources. The reports created under the previous HABHRCA provided solid background materials and served as a starting point. Federal agencies involved in research on marine, freshwater, and Great Lakes HABs and hypoxia contributed information from a wide range of resources, including scientific literature, project progress reports, and current and planned research and accomplishments. For further information on specific agency programs on HABs and hypoxia, see Appendix 3.

Table 1. Federal Departments and Agencies Comprising the IWG-HABHRCA. A key facet of the HABHRCA legislation is interagency collaboration. The IWG-HABHRCA is co-chaired by NOAA and EPA and is well-represented by a number of other Federal agencies with various interests and missions related to HABs and hypoxia. The IWG-HABHRCA has divided its work into four subgroups (HABs, Hypoxia, Engagement, and Great Lakes), each of which is headed by a different agency

Department of Agriculture
Department of Commerce*
Department of Defense
Department of Health and Human Services
Department of the Interior
National Aeronautics and Space Administration
Environmental Protection Agency*
National Science Foundation

* Denotes co-chairmanship.

Additionally, stakeholder engagement—specifically mandated by the 2014 reauthorization of HABHRCA—represented a key component of the IWG-HABHRCA's efforts in drafting this report and carrying out its other functions. This report summarizes the IWG-HABHRCA's efforts to interact with a wide range of interest groups, including relevant management and planning bodies, resource officials, economists, agricultural groups, tribal resource-management officials, scientists and public-health experts, industries affected by HABs and hypoxia, nonprofit groups, and the public. This report also offers recommendations for action that outline a path forward in effectively preventing and managing HABs and hypoxia events.

2. Engagement

In the process of developing this report, the IWG-HABHRCA engaged with approximately 1,000 stakeholders representing a wide range of interests, over 100 of whom submitted detailed input for consideration. Stakeholder representation included state coastal-management and planning officials; tribal resource-management officials; water- and watershed-management officials from both coastal and noncoastal states with source waters that drain into water bodies affected by HABs and hypoxia; national park and Federal land managers; public health officials; emergency management officials; science and technology development institutions; economists; industries and businesses affected by marine and freshwater HABs and hypoxia events; scientists with expertise concerning HABs or hypoxia from academic or research institutions; and numerous other stakeholders, including members of the general public. Moving forward, the IWG-HABHRCA will expand engagement efforts to include other relevant groups such as agricultural producers, fishermen, and watershed-advocacy groups.

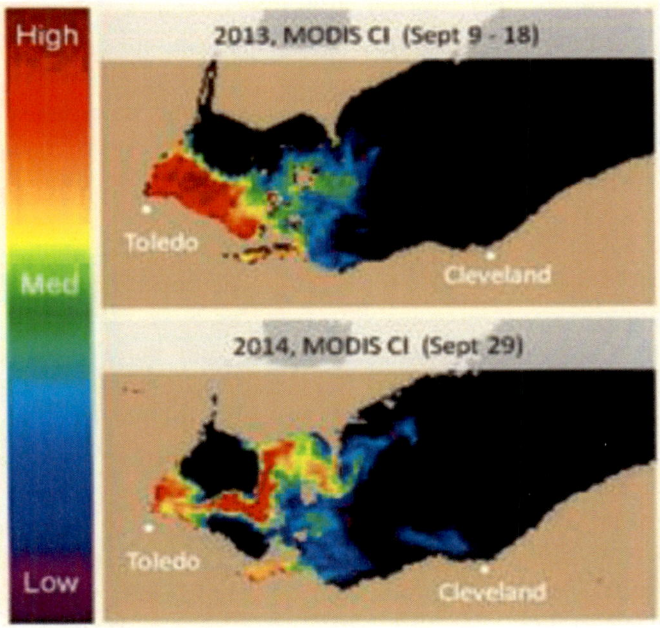

Figure 2.1. Example image of NOAA's Lake Erie HAB tracker, a tool that combines remote sensing, monitoring, and modeling to produce daily 5-day forecasts of bloom transport and concentration. This product takes daily satellite imagery and real-time monitoring to estimate the current expanse and intensity of the bloom to assist in predicting where the bloom will travel and at what concentrations. These predictions can provide water intake managers timely information for making public health decisions.

Stakeholder engagement occurred primarily through a series of five widely-advertised nationally- and regionally-focused webinars, during which participants had opportunities to interact with Federal representatives and ask questions before, during, and after each webinar presentation. An in-person stakeholder meeting co-funded by NOAA and the National Science Foundation (NSF) was held on April 14, 2015, at Bowling Green State University, OH, to discuss stakeholder needs specific to the Great Lakes region. Stakeholder engagement discussions related to HABs and hypoxia were held at conferences including Global Solutions to Regional Problems at Bowling Green State University, meetings and conference calls of opportunity (e.g., monthly calls of the National HAB Committee, a committee comprised of representatives from Federal and academic communities; the NOAA Ecological Forecasting Roadmap Meeting), and through direct engagement with agency-specific stakeholders (e.g., at the Interstate Shellfish Sanitation Conference). Finally, the IWG-HABHRCA established a specific email address for stakeholders to use to submit feedback: IWG-HABHRCA@noaa.gov.

Information submitted by stakeholders and through notes taken during the engagement activities were compiled, reviewed, and incorporated into this report as appropriate. Many contributions regarding the needs and challenges for mitigating, managing, and responding to HABs and hypoxia reiterated those contributed by Federal agencies. There were also numerous novel insights and ideas shared that were considered important and feasible for implementation.

A number of common themes emerged from the IWG-HABHRCA's conversations and meetings. Stakeholders specifically expressed the need for improved communication between all groups, including better, faster, and more consistent messaging about threats; and better coordination and information-sharing among relevant groups. During discussions with stakeholders, the IWG-HABHRCA realized that in many areas around the country, tools and information already exist that could benefit stakeholders; however, such tools and information are not used or communicated effectively. For example, the IWGHABHRCA received many questions and comments about a need for local and national citizen-science programs. Many individuals were unaware of the existence of nationwide opportunities such as the Phytoplankton Monitoring Network (PMN), or more localized examples like the Lake Champlain case study described in the next section. The stakeholder-engagement efforts described above gave the IWGHABHRCA opportunities to disseminate some of this information and connect groups with shared interests in HABs and hypoxia. It became clear nonetheless that a key element to addressing immediate and long-term local and national impacts of HABs and hypoxia is by improving when and how Federal agencies communicate with stakeholders and the public. Continuing to engage with stakeholders and maintaining communication pathways for research and resource sharing will be critical.

In addition, stakeholders emphasized a need for platforms and strategies to communicate the seriousness of HAB and hypoxia events. Many stakeholders said that it was difficult to communicate the severity or potential severity of HAB or hypoxia impacts without standard, consistent messaging. Social media was often cited as a way to facilitate communication; however, many people also indicated that the scope of reach of social-media platforms such as Facebook or Twitter is limited, especially with reaching certain populations or when attempting to effectively communicate the effects—not just the occurrence—of a HAB or hypoxia event. Stakeholders appreciated that there is a necessary balance of communication that has yet to be achieved. For example, providing too much information may lead to messages being ignored or rendered ineffective, whereas too little or infrequent messaging may result in messages being missed or misunderstood. IWG-HABHRCA engagement showed a clear need for inexpensive and more-widely available and used methods of communication in order to maximize the benefit of information provided to the public. Stakeholders also indicated that early warnings and predictions of HABs and hypoxia events are among the most important tools available for helping communities, regions, and the country prepare for and mitigate any potential effects, including by making better decisions related to resource and funding management, and preventing human- and animal-health impacts.

The extensive 2015 *Pseudo-nitzschia* bloom along the west coast of the United States demonstrates the importance of improving public engagement by providing early warnings. The areas affected by this bloom are home to a number of commercially and socially important species, including razor clams and crabs. The closure of the razor clam fishery in Washington State in May 2015 alone resulted in an estimated $9.2 million in lost income. Advance knowledge allowed communities to find alternative pathways for acquiring resources that this income typically helps fund, including school supplies used by children in indigenous communities. Additionally, providing early information helped inform the public about potential dangers of *Pseudo-nitzschia*, and act accordingly—the early warning triggered the closure of a fishery, thereby preventing human illness and death. Providing communities with early warnings is essential in allowing sufficient time for planning, thereby reducing the short- and long-term impacts that HABs and hypoxia events may have on social and economic health.

2.1. Examples of Future IWG and Agency Engagement

The IWG-HABHRCA plans to build on engagement to date by continuing to interact with stakeholders. Subsequent work products and progress reports will provide future opportunities to capture further stakeholder input. The IWG-HABHRCA and the individual agencies comprising it will also strive to leverage existing organizations and efforts aimed at informing or engaging specific stakeholder groups.

For example, NOAA has significant infrastructure, resources, and stakeholder relationships that could be leveraged better in the engagement process, thus potentially helping to meet needs expressed by stakeholders. USDA has also worked to expand citizen engagement. A team of Federal and university researchers and agricultural extension educators from the 12 States along the Mississippi and Ohio Rivers, known as the Southern Extension and Research Activities Committee number 46 (SERA-46), has partnered with the HABHRCA-mandated Mississippi River/Gulf of Mexico Watershed Nutrient Task Force (the "Hypoxia Task Force") and its members to increase efforts to manage complex natural-resource management issues. The group is working to expand and encourage the use of science-based nutrient management and other practices that help to reduce nutrient losses. SERA-46 is working to identify opportunities for states to share information, while also creating a network of leaders, including farmers, who strategize about agriculturally based nutrient losses.

Likewise, the Office of Water of the EPA has developed a series of webinars, outreach materials, and online resources to build awareness of HABs and nutrient pollution, and to increase communication on the impacts HABs have on human health, ecosystems, and the economy. The efforts include building an online database to share information; tours of HAB sites for resource managers; and websites containing information about HAB-related research efforts as well as available policies and guidelines for HAB prevention, control, and mitigation.

2.2. Case Study – Lake Champlain

Vermont and New York are home to one of the country's largest and best-known bodies of freshwater, Lake Champlain. The State of Vermont created the Lake Champlain Cyanobacteria Monitoring Program in response to the lake's first HAB event, in 1999. The program is unique in that it combines qualitative and quantitative sampling and observation information from citizen scientists, water suppliers, and state staff for HAB monitoring. The program sends out a weekly email to stakeholders in New York, Vermont, and Québec with relevant information and also posts data on the Vermont Department of Health's cyanobacteria website. The program has been highly successful, monitoring 87 locations around Lake Champlain and elsewhere in Vermont during 2014, and helping to protect boaters, swimmers, and other people who visit the lakes, as well as the region's drinking water and the many businesses that depend on tourists to stimulate the economy.

During webinars and in email correspondence with members of the IWG-HABHRCA, representatives of the program made it clear that social media was one of the most helpful tools for disseminating information to the public. The members of the program have also indicated that there is a need for coordinated messaging related to HABs among local, state, and Federal authorities. They stated that their ability to reach and engage lake users within their region is in part attributable to the fact that they have been able to communicate easily with their stakeholders and partners using consistent language and easily understood messaging.

3. Causes of HABs and Hypoxia

HABs and hypoxia are caused by several factors acting together, and indeed, the two types of events are often linked. HABs are created by a small subset of naturally occurring microscopic or larger, plant-like cyanobacteria or algal species, including diatoms and dinoflagellates. Under certain conditions, these species can form large masses, or "blooms." The algae in these blooms is ultimately consumed (typically by zooplankton) and recycled as fecal pellets, or dies naturally and sinks. Bacteria decompose these algal remains, a process which consumes dissolved oxygen contained in the water. When the rate of oxygen consumption in aquatic environments exceeds resupply, oxygen concentrations can quickly decline to levels insufficient to sustain most animal life, producing hypoxic conditions. Algal blooms also reduce water clarity and prevent sunlight from penetrating into the water and reaching submerged aquatic vegetation and benthic (seafloor) microalgae, causing them to release less oxygen that would normally help to replenish the water's oxygen supply. In these ways, HABs can contribute to hypoxia (CENR, 2010).

Physical, chemical, and biological conditions also influence the development and persistence of HABs and hypoxia. Such factors include lake morphology, water-circulation patterns, light, temperature, grazing pressure from plant-eating fish, viruses, and microbial mechanisms (Kudela et al., 2015). Most hypoxia caused or exacerbated by human activities occurs in waters that are susceptible to hypoxia due to the physical structure of the body of water in question. In many temperate lakes, estuaries, and near-coastal waters, a vertical density gradient, or stratification, occurs in summer months. (In freshwater, the density gradient is driven by temperature; in estuaries and coastal areas, it is driven by a combination of temperature and salinity.) This density gradient isolates a layer of bottom water and sediments from what is usually a well-oxygenated surface layer. Density stratification of the water column can encourage the growth of some species, like dinoflagellates or cyanobacteria, that commonly trigger HABs, because of the organisms' ability to move vertically and optimize their access to light and nutrients. Stratification also reduces the rate at which oxygen can replenish in deeper waters, meaning that decomposition of organic matter in bottom waters and sediments can rapidly deplete available oxygen in stratified bodies of water. Nutrient pollution that stimulates algal blooms can result in much more organic matter reaching bottom waters, thus driving hypoxic conditions to much more severe levels than would occur naturally.

Hypoxia is not limited to stratified waters, however. Episodic hypoxia can occur in unstratified and shallow waters in areas of extreme algal production caused by nutrient pollution. In these increasingly common situations, algal respiration at night can lower oxygen to near zero. Because this type of hypoxia often occurs in shallow habitats that are preferred by fish populations, it can, in a matter of hours, trap fish and lead to large fish kills.

Hypoxia can also occur in near-coastal waters due to upwelling. Upwelling is the process by which warm, less-dense surface water is drawn away from along a shore by offshore currents and replaced by cold, denser water brought up from the subsurface. Upwelling of offshore ocean-bottom waters that are permanently depleted of oxygen and lead to hypoxia. This phenomenon occurs due to strong and prolonged winds from a favorable direction, and it most commonly occurs in the Pacific Northwest.

While natural processes may cause HABs and hypoxia, they cannot completely account for the observed marked increase in the number and duration of HABs and hypoxia events around the world. The incidence of hypoxia globally has increased tenfold over the past 50 years, and by almost thirtyfold in the United States since 1960, with more than 300 aquatic systems recently experiencing hypoxia (Diaz and Rosenberg, 2008; CENR, 2010). Every state in the United States now experiences some kind of HAB or hypoxia event, in many cases annually, and previously unrecognized HAB species have emerged in some locations (Appendix 1).

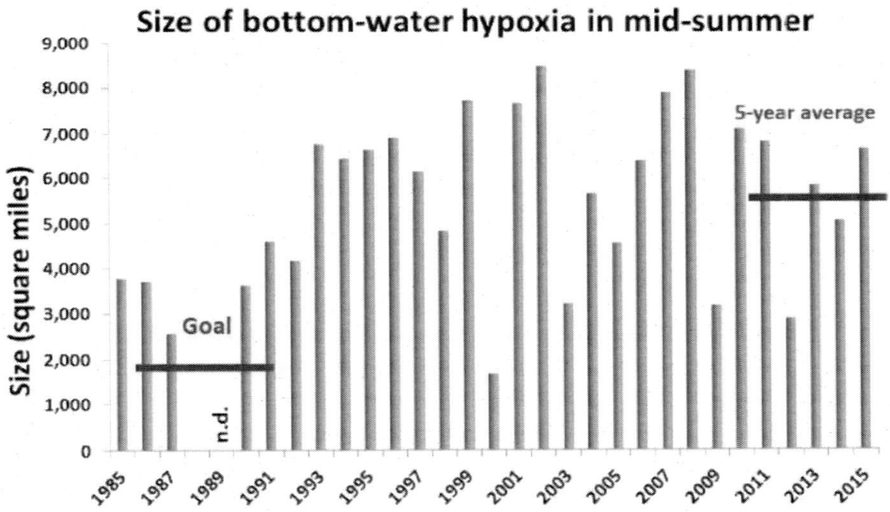

Figure 3.1. Size of Bottom-Water Hypoxia in Mid-Summer, Gulf of Mexico. This figure demonstrates the discrepancy between the current level of bottom-water hypoxia in the Gulf of Mexico, an annual occurrence, and the goal level. (Note that bottom-water hypoxia is not the full measure of the hypoxic zone.) "Goal" refers to the Mississippi River/Gulf of Mexico Watershed Nutrient Task Force's (the "Hypoxia Task Force") objective of reducing the area of the Gulf of Mexico hypoxic zone to less than 5,000 km^2 (approximately 1,930 square miles, as shown above). The Task Force also agreed to reduce the nutrient load by 20 percent by the year 2025, as a milestone toward reducing the hypoxic zone to less than 5,000 km^2 by 2035. (Data Sources: Nancy N. Rabalais, L, and R. Eugene Turner, Louisiana State University; Funding Sources: NOAA Center for Sponsored Coastal Ocean Research and EPA Gulf of Mexico Program).

This increase is largely driven by ecological changes, food-web alterations, and the introduction of HAB species into new regions due to international commerce and water-flow modifications. Another important driver of HABs and hypoxia events is the export of large quantities of nutrients, such as phosphorus and nitrogen, and organic matter into coastal waters in areas of high population density or near developed watersheds (CENR, 2010). Excess nutrients and organic matter can stimulate harmful algal growth. Nutrient pollution can come from both direct ("point") and indirect ("nonpoint") sources, including agriculture; municipal, and industrial wastewater; urban and suburban stormwater runoff; and aquaculture. In addition, atmospheric pollutants—especially nitrogen from fossil-fuel combustion, volatilization of fertilizer and animal waste, and industrial outputs—can be deposited from the air to watersheds or directly into water, leading to increased nutrient levels. Overall, the greater the input of nutrients and organic matter into a body of water, then the greater the chance of hypoxic conditions developing and continuing.

The Northern Gulf of Mexico Hypoxic Zone

The northern Gulf of Mexico hypoxic zone is possibly the most widely known instance of hypoxia. It is described in detail in the Mississippi River/Gulf of Mexico Watershed Nutrient Task Force 2015 Report to Congress. The size of this hypoxic zone varies considerably from year to year, depending on natural and anthropogenic factors (Figure 2.1). The 2015 hypoxic zone measured 6,474 square miles as of July 28-August 3, approximately the size of Connecticut and Rhode Island combined (NOAA, 2015). Potential climate effects—including frequent, severe weather events; prolonged drought; and early spring runoff— compound this variability within the reassessment timeline.

(*See "The Mississippi River/Gulf of Mexico Watershed Nutrient Task Force 2015 Report to Congress" for further details.*)

3.1. Effects of Climate Change on HABs and Hypoxia

There is not a simple overarching relationship among climate change, HABs, and hypoxia. The character of HABs and hypoxia events can be affected in a variety of ways, including by increasing water stratification, altering the underwater-light environment, increasing dissolved carbon dioxide levels, decreasing aquatic pH, changing circulation patterns, altering nutrient availability, and modifying the rest of the biological community (Wells et al.,

2015). Certain aspects of climate shifts—including changes in extreme-weather events, precipitation timing, and climate variability—are projected to increase anthropogenic sources of nutrients reaching water bodies, and exacerbating the impacts of HABs and hypoxia (Rabotyagov et al., 2014). This effect is likely to be compounded as increased human activity further elevates the amount of nutrient inputs. Climate change also leads to increases in freshwater temperature and stratification, increasing the likelihood of HABs and hypoxia events (Gowen et al., 2012; Paerl and Paul, 2014; Beaver et al., 2014; Wells et al. 2015).

The response of low-biomass HABs (and some high-biomass HABs) to climate change or increasing nutrient inputs is much more difficult to predict (Anderson et al., 2012; Bresnan et al. 2013; Davidson et al., 2013; Gowen et al., 2012; Wells et al. 2015). These HABs include many of the toxic marine and freshwater species that threaten human health. The difficulty arises because growth and toxicity of each species and strain responds differently to changing environmental conditions, and because each is a part of a larger plankton community. Studies of the impact of climate change on algal toxicity have demonstrated variable results on toxicological outcomes, either increasing or decreasing toxicity (Griffis et al., 2013), although research has also shown direct correlations between changing climate and a higher prevalence of human-health syndromes such as ciguatera fish poisoning (CFP) in some regions (Gingold et al., 2014).

Scientists believe, however, that climate change will result overall in net increases in HABs and hypoxia events, and that the spatial and temporal patterns of these events will change (CENR, 2010; Griffis et al. 2013). For example, scientists hypothesize that, with continued shifts in temperatures and ocean patterns off the United States west coast, HABs causing paralytic shellfish poisoning in Puget Sound, Washington, will occur earlier and last longer, meaning that shellfish-harvesting closures imposed to protect human health will need to be extended (Moore et al., 2011; Melillo, 2014).

Better assessment of the current and projected future impacts of climate will require long-term, retrospective records of phytoplankton and environmental conditions at specific locations, combined with observing systems to monitor future changes (Wells et al., 2015).

4. THREATS CAUSED BY HABS AND HYPOXIA

As concerns about HABs increased in the 1990s, the Federal Government began to devote greater attention to these issues (U.S. Senate Report 105-357; 1998). Since that time, under the direction of HABHRCA and guidance of National Plans (e.g., *Marine Biotoxins and Harmful Algae: A National Plan*; *Harmful Algal Research & Response National Environmental Science Strategy 2005-2015*), scientists understand much more about the impacts of HABs and hypoxia events.

4.1. Ecosystem Impacts

The mechanisms by which HABs cause harm vary. While most algal species are not toxic (indeed, algal species are widely recognized as one of the cornerstones to life on Earth, and form the basis of the food chain as food for zooplankton and fish), there is a small but important subset of algal species that are known to be toxic (see Appendix 1). HABs caused by toxic algal species can kill fish or shellfish directly, and can sicken birds, marine mammals, and people who drink contaminated water and eat contaminated seafood if sufficient amounts of HAB toxins are ingested.

The mortality or relocation of algae-eating aquatic life can lead to or exacerbate the emergence of HABs, allowing the HABs to take over and dominate ecosystems. Such circumstances may occur independently of the effects of HABs and hypoxia (e.g., when coral bleaching leads to dead or dying coral reefs), or may be linked to these effects in a positive feedback loop that further reinforces hypoxic conditions (e.g., when hypoxia in an aquatic ecosystem reduces the populations of algae-controlling fish species, allowing algal blooms to continue unchecked) (Rosenblatt et al., 2013). Hypoxia-induced changes in sediment chemistry can also lead to nutrient release, further promoting HABs, especially in phosphorus-limited freshwater systems.

Hypoxia alone is known to have a wide range of detrimental impacts on aquatic species. Low concentrations of dissolved oxygen and toxic compounds can be lethal to aquatic species, and increases in hypoxia and HAB events have led to increased frequencies and magnitudes of fish kills (Thurston, 2002; Thronson and Quigg, 2008).

HABs are an indicator of environmental degradation and are often associated with reduced biodiversity and greater environmental instability. Exposure to toxins released by select HABs can affect physiology, pathology, behavior, and reproduction rates of organisms (Baganz et al., 1998; Greenfield et al., 2014). The direct and indirect effects of HABs and hypoxia can disrupt entire food webs, with extensive and varied impacts. In a coral-reef food web, for example, the coral reefs provide food sources and habitats for commercially important fish species. Increasingly, however, excess nutrients—in conjunction with overfishing and ocean-temperature changes—are decimating corals faster than the corals can adapt for survival. Nutrients promote the growth of harmful macroalgae that overgrow and kill corals. Normally, herbivorous fish eat these macroalgae, allowing space for larval corals to settle and regrow. Overfishing reduces the numbers of these fish such that the fish populations cannot control the overgrowth of the HABs (Rosenblatt et al., 2013). Exceedingly high summertime temperatures also are causing massive coral bleaching (mortality) events. These mass-mortality events create open space on the reef ideal for colonization by the faster-growing HABs, resulting in a further loss of viable coral habitat. Because coral reefs act as physical barriers that serve as coastline stabilizers and storm-surge buffers, coral loss can have catastrophic effects on ecosystems and coastal communities. HABs have already damaged coral reefs that play an important role in dampening coastal storm impacts in the United States (HARRNESS, 2005; Riegl et al., 2012).

HABs also adversely affect other shoreline-stabilizing organisms. Dense HABs can directly inhibit growth of beneficial vegetation, such as seagrasses, that often serve as storm buffers, by blocking sunlight penetration into the water column and by fouling beaches. HABs harm shoreline-stabilizing oysters by inhibiting growth and reproductive rates, and ability to fight off disease and other stressors (Lopez et al., 2008b).

HABs and hypoxia can also have non-lethal effects, including shifts in spatial distribution of organisms, changes in community structure caused by migration of fish and other mobile organisms like shellfish, and alteration or blockage of normal migration routes of fish and invertebrates (Craig, 2012; Craig and Bosman, 2012; Roman et al., 2012; Kraus et al., 2015). These effects can alter the availability of fish and other organisms that are important either as sources of food or sources income generated by industries such as fishing or tourism. While the specific concentrations (Figure 3.1) of low dissolved oxygen that affect animals vary by species, effects generally appear when oxygen levels drop below about 3 milligrams/liter (Diaz and Rosenberg, 1995; Ritter and Montagna, 1999; Breitburg et al., 2001; Rabalais et al., 2001; Karlson et al., 2002).

4.2. Human and Animal Health Impacts

HABs can adversely affect the health of all animals and some plants, with the potential to seriously harm or kill thousands of organisms every time blooms occur. Indeed, sick or dying animals are often the first indicators of a toxic bloom and may serve as harbingers for other potential harmful effects. HAB-related deaths of marine animals, including manatees, seals, whales, sea turtles, and sea birds, can involve hundreds of individuals during each event (NOAA, 2004). HABs have been the most common cause of unusual mortality events (UMEs) in marine mammals over the last 20 years (Gulland and Hall, 2007; Simeone et al., 2015). Blooms commonly cause fish kills and bird mass-mortality events, and can sometimes poison pet dogs and livestock.

It is less well-known whether, and to what degree, the exposures to HAB toxins described above cause long-term health effects in humans, such as liver disease, cardiovascular disease, developmental defects, or neurobehavioral illness. Additional research into the effects of exposure to HAB toxins will help to further describe the acute and longer-term health risks to humans and other animals. Humans and animals may be exposed to algal toxins via consumption of contaminated seafood, dietary supplements derived from cyanobacteria, or foods grown with contaminated irrigation water. Exposure may also occur through drinking water, inhalation, skin contact, or accidental ingestion of water while engaging in recreational activities. A summary of acute or short-term human health consequences of HAB toxin exposure can be found in Appendix 2.

Even blooms that are not toxic can cause damage by forming masses that suffocate fish by clogging their gills, block light from bottom-dwelling plants, (Lopez et al., 2008a, 2008b; Delegrange et al., 2015; Kudela et al., 2015), clog the water-supply intakes in water-treatment plants (Caron et al., 2010), and adversely alter food webs by displacing beneficial species. (Lopez et al., 2008b; Delegrange et al., 2015). When HABs lead to hypoxia, they can stress or kill many organisms, including fish and shellfish, their predators, and the prey they depend on. While most mobile organisms are able to avoid hypoxic waters by swimming or crawling away, organisms that are unable to move or are otherwise trapped within an hypoxic zone become physiologically stressed and can die if exposure is prolonged or severe (CENR, 2010).

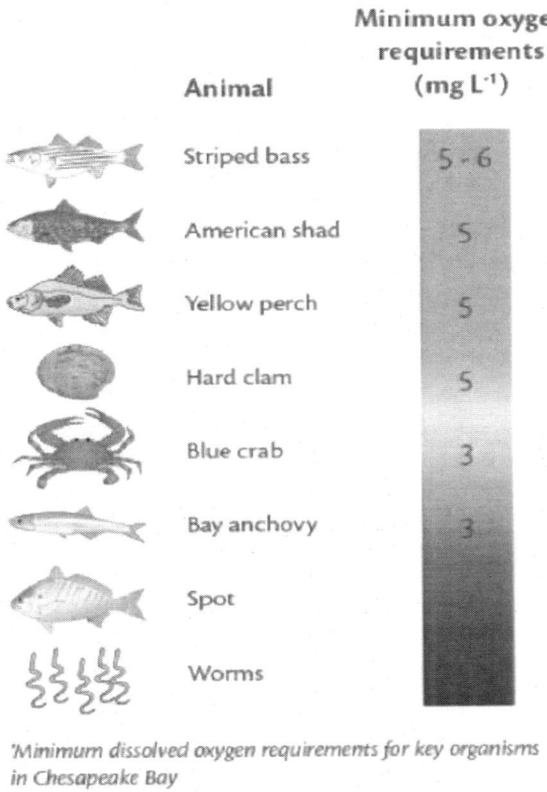

Figure 4.1. Minimum Oxygen Requirements Chart (Source: 2010 Hypoxia Scientific Assessment).

Hypoxia means "low oxygen". In aquatic and marine systems, low oxygen generally refers to a dissolved oxygen concentration less than 2 to 3 milligrams of oxygen per liter of water (mg/L), but sensitive organisms can be affected at higher thresholds (4.5 mg/L). A complete lack of oxygen is called anoxia.

Hypoxic waters generally do not have enough oxygen to support fish and other aquatic animals, and are sometimes called dead zones because the only organisms that can live there are microbes. The criteria set for health of various species in Chesapeake Bay are a good example of how one definition for hypoxia is not possible (EPA, 2003).

4.2.1. Examples of Seafood Poisoning

HAB toxins are commonly passed through aquatic food webs: from algae to filter-feeders to fish to top predators, including people. Through this process, the toxins may accumulate to levels that can cause various poisoning syndromes, which in turn cause illness or even death in humans or other consumers (Lewitus et al., 2012; reviewed in Munday, 2014).

The shellfish-poisoning syndromes are known as paralytic (PSP), diarrhetic (DSP), neurotoxic (NSP), azaspiracid (AZP), and amnesic (ASP) shellfish poisoning. Except for ASP, which is produced by certain diatoms (Bates et al., 1989), all are caused by biotoxins synthesized by a class of marine algae called dinoflagellates. Significantly, DSP appears to be an emerging threat, with shellfish-harvesting closures as a result of DSP reported in 2008 off the coast of Texas (Deeds et al., 2010), and other DSP events occurring in 2011 in Washington (Trainer et al., 2013) and British Columbia (Taylor et al., 2013). Dealing with this newly emerging group of toxins is a current challenge for several impacted state programs. Another prominent human syndrome, CFP, is caused when humans consume fish contaminated with accumulated ciguatoxins, the precursors of which are produced by microalgae that grow on seaweeds and other surfaces in coral-reef communities (Lehane and Lewis, 2000).

Agencies work to track, reduce, and prevent HABrelated illness

The Centers for Disease Control and Prevention developed a HABs module for the National Outbreak Reporting System (NORS) to accumulate clinical data on people and animals poisoned by HABs. This system has recorded HAB-associated human illness after recreational water exposure in the last few years (Yoder et al., 2004; Dziuban et al., 2006; Hilborn et al., 2014). Reports described gastrointestinal, dermal, and respiratory signs and symptoms. No algal bloom-associated waterborne disease outbreaks associated with drinking water have been reported through NORS to date.

4.2.2. Examples of Water and Air Poisoning

HAB toxins can be released directly into the water or air, either naturally through processes like algal cells breaking up and releasing toxins due to water turbulence, or through human activities such as water treatment or power boating. For instance, in the case of one of the most well-known types of HAB events, toxic red tides in the Gulf of Mexico (*Karenia brevis*), HABs release toxins into sea-spray aerosols. These aerosols are particularly problematic at beaches, as they can cause respiratory distress to local residents, lifeguards, and beachgoers. Inhalation of these aerosols leads to respiratory irritation, coughing, and other symptoms (Backer, 2009). Exposure from HAB toxins released into the air above fresh waters is a potential problem, but is less-well characterized. Most health effects associated with inhalational exposure to freshwater HABs have occurred among those recreating on or in HAB contaminated water (Giannuzzi et al., 2011; Hilborn et al., 2014).

4.2.3. Complicating Factors

One complicating factor in assessing the health impacts of HABs is the fact that not all toxin-producing algae always release their toxins into surface waters. For example, anatoxin-a and microcystin toxins tend to be found intracellularly and are only released when algal cells rupture or die. Conversely, a large amount of cylindrospermopsin is released into the water by algae that are living. Hence the presence of toxin-producing algal species is not always independently indicative of the extent of human and animal health risks. As resource managers work to treat water effectively, they are challenged to consider detailed information such as algal growth patterns, environmental conditions, dominant species of algae, and the toxicological properties of relevant compounds.

Furthermore, in addition to directly inducing illness, HAB toxins can interact with other internal and external agents to alter the course of disease. Microcystins, a type of toxin produced by some HABforming cyanobacteria, have been shown to initiate and promote liver cancer in laboratory animals (Fujiki and Suganuma, 2011), and epidemiological studies in China and Serbia suggest that microcystins may have contributing roles in the development of human primary liver cancer (World Health Organization, 2008; Svircev et al., 2009; Tian et al., 2013).

4.3. Socioeconomic Impacts

HABs and hypoxia also may have serious effects on a community's social and economic health. It is important for cost-benefit analyses and risk assessments to include a clear explanation and reliable estimate of the economic and social costs of HAB and hypoxia events, as well as a discussion of the populations that these costs affect most. Social impacts of HABs and hypoxia events are generally much more difficult to quantify than economic impacts, although efforts are underway to address this challenge. Ongoing studies through NOAA and the National Institutes of Health's National Institute of Environmental Health Sciences (NIEHS) to identify animal and human populations at higher risk for exposure to and adverse effects from

HABS and hypoxia events, for instance, could support improved analysis of social impacts.

Social Impacts of HABs

In August 2014, Lake Erie experienced a bloom near the intake to the drinking water treatment plant serving the city of Toledo, Ohio. Toledo issued a "do not drink or boil advisory" that affected nearly 500,000 customers due to the presence of cyanotoxin that exceeded the safe drinking water threshold recommended by the World Health Organization (WHO) (Wilson 2014).

Although no national estimate of the economic costs of HABs has been conducted, coastal and inland HAB events in the United States have resulted in millions of dollars in costs to communities, including losses in income from commercial fishing, recreation, and tourism industries, public health costs of illness, and expenses for monitoring and management (Lopez et al., 2008a, 2008b). A 2015 EPA report evaluated the costs associated with controlling nutrients at their sources, as well as those associated with the impacts of nutrient pollution for a number of economic sectors, with a focus on HABs and hypoxia. The report found that in Ohio, a persistent HAB in a lake caused communities to lose between $37 million and $47 million in tourism dollars between 2009 and 2010. Another HAB off the coast of Maine led to $2.5 million in losses of soft shell clam harvests from April to August 2005 (EPA, 2015a). It should be noted, however, that there is dissention within the scientific community of the methodologies used to derive cost estimates for coastal HABs (Hoagland and Scatasta, 2006; Grimes et al., 2012; EPA, 2015a).

These costs can manifest in a variety of ways—including public-health costs, commercial-fisheries losses, recreation and tourism losses, and monitoring and management costs— and can differ between freshwater and coastal HABs. Freshwater HABs can impose substantial costs when such events contaminate sources of drinking water, leading to high public-health costs as a result of water testing and treatment, provision of alternative drinking-water sources, and loss of municipal revenue due to water-use bans. HABs and hypoxia can also affect food and drug production, with associated loss of revenue. Toxic HABs pose a potential hazard for dietary supplements containing cyanobacteria, meaning that government agencies (such as the Food and Drug Administration) and private companies have to develop and use costly detection techniques to prevent toxins from entering the commercial market. Livestock can die after drinking water contaminated by HAB toxins, and algal blooms may decrease fish production, cause off-flavor and objectionable odors in fish, and lead to fish kills because of oxygen depletion. Indeed, hypoxia can have large economic impacts on the fishing industry, through fish mortality, reduced growth, and reduced fish reproduction, as well as by increasing the time needed to fish to compensate for decreased fish availability. Hypoxia has been linked to the collapse or impairment of a number of commercially important fisheries worldwide (Mee, 2006); it can be challenging, however, to effectively separate the effects of hypoxia from the effects of overfishing (Breitburg et al., 2009). It is likely that the two causes are linked: that hypoxia decreases productivity and resilience of exploited populations, making them more vulnerable to collapse in the face of heavy fishing pressure, as was evidenced in a study of Gulf of Mexico brown shrimp (Smith et al., 2014).

In coastal areas, excess nutrients and warmer water temperatures can promote large offshore macroalgal (seaweed) blooms (Figure 3). These blooms are driven onto shore by prevailing currents and winds, particularly during storms. When this happens, beaches, shorefront properties, and marinas can become inundated in rotting seaweed, which can be three or more feet deep. This huge accumulation of biomass causes myriad problems, including fouling small-boat motors, deterring tourism, and degrading property values. The cleanup and disposal costs associated with these blooms constitute a significant economic loss which must be borne by local governments and individual property owners. These costs have risen over time as macroalgal blooms have increased in intensity and frequency (Lapointe et al., 2004).

In addition to imposing economic impacts, HABs and hypoxia can disrupt coastal communities socially and culturally; for instance, by harming fisheries in communities that rely on subsistence fishing as a food source and livelihood, as is the case for Native American communities around the country (Lewitus et al., 2012). HABs and hypoxia also disrupt other ecosystem services[1] that underpin social and cultural health (Lipton and Hicks, 1999, 2003; Bricker et al., 2006), although these effects are complex and can be difficult to fully capture and quantify.

5. Current Management Actions for HABs and Hypoxia: Successes since 2004

Through the 1970s, organic-matter loading from sewage and industrial discharge was a primary cause of eutrophication in the United States, leading to HABs and hypoxia events (Diaz and Rosenberg, 2011; Pellerin et al., 2014). Actions following the passage of the Clean Water Act (CWA) in 1972 reduced point-source organic loading and, later, reduced point-source phosphorus and nitrogen loading, contributing to some improvements in hypoxia problems. With improvements in nutrient contributions for these point sources, nonpoint runoff of nutrients became a more significant component of water-quality deterioration in inland and coastal systems in the United States as development and agricultural production increased (Vitousek et al., 1997; Bricker et al. 1999, 2007; Diaz and Rosenberg, 2011; Pellerin et al., 2014). Recently, efforts have been made to address the proliferation of nonpoint sources of nutrient pollution, such as agricultural, urban, and suburban runoff through regulatory and voluntary means.

Effective management of HABs and hypoxia is critical to minimizing negative impacts on living resources; human health; and local, regional, and national economies. The status of various management approaches in the field was summarized in the reports produced per the 2004 HABHRCA reauthorization,[2] and the following general recommendations for how to improve HAB prediction and response were made:

- Conduct research focused on the development, demonstration, and technology transfer of methods for the prevention, control, and mitigation (PCM) of HABs and HAB impacts;
- Develop a coordinated, regionally organized HAB event response; and
- Ensure availability of core infrastructure to support HAB research and response.

The concepts developed through these plans—i.e., prevention, control, and mitigation—are key to managing HABs and hypoxia. *Prevention* refers to environmental-management actions taken to reduce the incidence and extent of HABs and hypoxia events. *Control* (or suppression) refers to strategies that directly kill HAB cells or destroy their toxins, physically remove cells and toxins from the water column, or limit cell growth and proliferation. These strategies aim to reduce the impacts of HABs and hypoxia events on people and commerce by targeting the immediate causative agents of these events. *Mitigation* refers to responding to an existing or ongoing bloom by taking steps to restrict, inhibit, or prevent associated undesirable impacts on the environment, human health, or human economies and communities. Prohibiting seafood that is contaminated with HAB toxins from entering commerce is an example of a management strategy to mitigate human-health impacts. Mitigation is the area of HAB management where the most immediate potential exists to reduce impacts, given that many such activities are already underway (Seltenrich, 2014).

The Federal agencies addressing critical components of threats posed by HABs and hypoxia have made many advances since the 2004 authorization of HABHRCA. In response to community and policymaker needs, NOAA has established a research program focused on preventing, controlling, and mitigating HABs. Since 2004, HAB- and hypoxia-forecasting systems are in place in several locations around the country, and agencies have developed more expansive and detailed monitoring and detection systems. These systems have enabled timely closing and reopening of shellfish beds, recreational areas, and drinking-water sources to avoid harmful impacts to people, marine mammals, and the economy. Federal efforts have helped advance understanding of nutrient-loading mechanisms, helping managers assess the efficacy of current nutrient-reduction practices and improve nutrient management and management programs. Additionally, the efforts have led to the development of effective strategies for modeling and mitigating hypoxia. Federal activities have also helped raise awareness of HABs and hypoxia among external stakeholder groups, most notably among groups outside of academic and research sectors, thereby facilitating collaborative, cross-sectoral efforts to reduce the impacts of HABs and hypoxia on humans and animals.

This section provides some examples of progress in HAB and hypoxia prevention, control, mitigation, and monitoring that has been made since 2004. Additional information about specific activities is contained in Appendices 4 and 5.

5.1. Prevention

5.1.1. Primary Prevention Strategies

Although regulating the factors known to contribute to the development of HABs and hypoxia is often difficult and not always feasible, certain contributing factors—including changes in land use, increased nutrient loadings, altered hydrology, introductions of new species, and increased aquaculture in HAB- and hypoxia-prone areas—can be regulated to some extent. Management strategies evolve as knowledge grows. Primary strategies for controlling HABs and hypoxia currently include the following:

- Minimizing land-based nutrients flowing into coastal and inland waters;
- Reducing nitrogen emissions in the airshed to reduce precipitation-related deposition of nitrogen compounds from the air to the land;
- Aerating systems and flushing dams to prevent hypoxia; and
- Reducing new introductions of algal species through genetic control (genetically engineering species that are purposely introduced into an ecosystem in order to alter the environmental tolerances, reproduction, or other processes of undesirable species) and environmental control (manipulation of an ecosystem environment, such as altering ecosystem habitat to disfavor growth of undesirable species.

5.1.2. Excess Nutrient Input

Developing effective strategies for preventing HABs and hypoxia requires a sufficient understanding of the causes of these issues. One cause that has attracted particular focus is excess nutrient input. Excess input of nutrients—including nitrogen, phosphorus, and legacy nutrients--is one of the primary causes of HABs and hypoxia events, and can exacerbate the development of HABs or hypoxia in systems that are already prone to these issues. There are several nutrient drivers of HABs and hypoxia that are expected to increase with the growing human population, including intensification of agriculture, conversion of land for development, and higher energy consumption that results in greater nitrogen air emissions. Aging water and sewage systems can also leak nutrients that influence the development of HABs and hypoxia (Paerl et al., 2011), and there is emerging research that discharge of submarine groundwater— that is, water located underneath soil and rock in marine ecosystems—may contribute to hypoxia in coastal waters (Moore, 2010; McCoy et al., 2011). The key to reducing nutrient drivers of HABs and hypoxia is to develop and implement cost-effective and scientifically sound nutrient-reduction strategies in rivers and streams, lakes and reservoirs, and estuaries and near-coastal marine waters.

Currently, there are efforts across the country to better understand the causes and consequences of excess nutrient input, including how complex interactions among groundwater systems, surface-water systems, biogeochemical processes, and other factors affect nutrient loading; nutrient transport through water systems (including evaluating changes in upstream ecosystems that can impair areas downstream); and how the contributions of atmospheric nitrogen and groundwater nutrients to the development of HABs and hypoxia. NOAA's predictive modeling for HABs and hypoxia uses an integrated approach, which consists of using satellite images; remote sensing buoys; a comprehensive monitoring program in Lake Erie and Saginaw Bay, Lake Huron; and advanced genetic techniques to understand the long and short-term seasonal dynamics of HAB events. The data is used to inform predictive models used by key Great Lakes stakeholder groups, such as drinking water managers.

The USDA's Conservation Effects Assessment Project (CEAP), for instance, examined the effects of existing conservation activities on water quality and soil resources in edge-of-field and large river basins—such as the Mississippi River Basin, Great Lakes, and Chesapeake Bay—and in small watersheds. USDA has also estimated needs for conservation treatments to reduce soil erosion and nutrient transport in vulnerable ecosystems. Projects like these provide information that communities, resource managers, and decision makers need to make sustainable water-resource decisions. As a part of its larger forest-restoration and sustainable-forestry programs, the USDA Forest Service is maintaining and promoting forested buffer strips along streams to reduce nutrients inputs and manage aquatic HABs and hypoxia.

Other efforts focus on enhancing continuous measurement of nutrient concentrations, which improves understanding of HABs and hypoxia by more accurately quantifying nutrient loads in water bodies, giving insight into processes that control nutrient variability in freshwaters, and providing real-time data for early detection of trends. The U.S. Geological Survey (USGS), in partnerships with other state and Federal agencies, is using advanced optical-sensor technology to accurately track nitrate levels on an hourly basis at 104 small streams and rivers nationwide, including more than 40 sites throughout the Mississippi River Basin. Better nutrient measurement also improves the accuracy of, and reduces the uncertainty in, models used to calculate nutrient loading in coastal and other aquatic environments. Because these models are typically based on relationships between streamflow and measurements of nutrient concentration, insufficient nutrient data can compromise model reliability, which is particularly problematic during extreme climatic periods. For example, examining nitrate data collected continuously in the Mississippi River basin during the drought-to-flood years of 2012-2014 showed that the concentration–discharge relationships for these events were not well-represented by historical data (Pellerin et al., 2014).

Progress on addressing nutrient pollution has occurred through the work of the Hypoxia Task Force and the creation of Annex 4 of the Great Lakes Water Quality Protocol of 2012, an agreement between the United States and Canada related to managing, protecting, and restoring the waters of the Great Lakes. Annex 4 specifically outlines an action plan for reducing nutrient inputs in order to reduce HABs (Great Lakes Water Quality Protocol, 2012). Additionally, in August 2015, President Obama signed into law the Drinking Water Protection Act (P.L. 114-45), which required EPA to develop and submit to Congress a plan evaluating the risk of harmful cyanotoxins to human health. The plan—"Strategy Submitted to Congress to Meet the Requirements of P.L. 114-45," produced in November 2015—includes steps and timelines for evaluating the human-health risk from drinking water contaminated by algal toxins; establishing a list of algal toxins that may have an adverse effect on human health; determining whether to publish health advisories for algal toxins; establishing guidance regarding feasible analytic methods, monitoring guidance, and treatment options (including source-water protection); and outlining a Federal strategy for entering into cooperative agreements with and provide technical assistance to affected states and public-water systems (EPA, 2015c). The Drinking Water Protection Act also directs EPA to assemble and publish information from each Federal agency that has examined or analyzed cyanobacteria or algal toxins, or addressed public health concerns related to harmful algal blooms, and identify information gaps.

5.2. Control

Chemical, biological, physical, and environmental controls can be used to regulate or suppress HABs and hypoxia (Jewett et al., 2008). Since all control strategies involve modifying the natural environment, it is necessary to carefully consider the positive and negative impacts of deploying a given control in a given setting. For example, controlling HABs using bloom suppression is only useful and economically feasible on small scales (e.g., aquaculture settings, small water bodies), where the environmental and economic impacts of the HABs greatly outweigh those of bloom-suppression control method. 5.3.

Mitigating and controlling the impacts of HAB toxins on human health

An important component of controlling and mitigating HABs is the ability to limit or extract toxins from drinking water and food sources. EPA's Office of Research and Development is exploring the use of an emerging "green" technology for removing microcystins from drinking water. The majority of drinking water treatment studies have focused on removing microcystins; few studies exist with other toxins, mixtures of toxins, and mixtures of cyanobacteria. Marine biotoxins increasingly will be a concern in areas where seawater is desalinated as a source of drinking water (Appendix 3).

5.3. Mitigation

Currently, HAB and hypoxia mitigation is more practically feasible than HAB and hypoxia prevention or control. Effective mitigation requires early warnings of the emergence or increased risk of HABs or hypoxia. Early warnings, in turn, require monitoring of the input of nutrients into coastal and inland waters, including monitoring both levels of nutrients in water bodies and in their primary source waters. Early warnings also depend on models that predict the likelihood of HAB or hypoxia development based on observed conditions. For instance, EPA has developed models that predict chlorophyll *a*, or "trophic state," for nearly all lakes in the United States in order to determine pollution levels and the likelihood of cyanobacteria presence in inland waters (Hollister et al., in review).

Early warnings allow managers to mitigate the health and economic impacts of HABs or hypoxia events by making preparations in advance, focusing sampling in critical areas, and issuing advisories warning the public of potential impending consequences. Mitigating the negative human-health impacts of HABs involves management strategies for ensuring that seafood contaminated with HAB toxins is identified and prevented from entering into interstate commerce. For instance, FDA produces a "Hazards Guide" to help with the early identification of HAB toxins in seafood, and to help seafood processors develop "hazard-analysis critical control-point" plans. These plans are designed to prevent seafood contaminated with HAB toxins from reaching consumers. One example of mitigation of economic impacts comes from the State of Washington, which in spring 2015 issued early warnings of domoic acid contamination in razor clams, caused by an outbreak of *Pseudo-nitzschia* on the west coast of the United States. These early warnings led to and allowed for selective beach openings, saving at least $3 million per year for the State's coastal fisheries by providing more accurate and real-time information on the presence of the algae and toxin (Dyson et al., 2010).

Federal agencies have made some important progress in HAB and hypoxia mitigation, including limiting impacts on human health. For instance, scientists in the FDA's Center for Food Safety and Applied Nutrition have identified and characterized HAB toxins in seafood and developed analytical methods for monitoring and controlling these toxins. The FDA has made critical advances in food safety by identifying the additional fish species and regions where ciguatera exists and updating guidance to industry accordingly. Such guidance is used to develop controls to prevent these toxins from entering the Nation's food supply. FDA has also implemented programs that strengthen partnerships in the public and private sector for identifying sporadically-occurring, novel, and emerging toxins, as well as vulnerable seafood products. For example, methods have been developed and validated recently for detecting microcystins in blue-green algal supplements, thereby providing tools for regulatory enforcement of safe, commercially available cyanobacteria-derived dietary supplements. Similarly, researchers supported by the NIEHS detected the cyanobacterial toxin beta-methylamine alanine (BMAA) in certain aquatic animals like crabs and shrimp that are consumed by humans. BMAA is a toxin linked to neurodegenerative diseases in humans, such as amyotrophic lateral sclerosis (more commonly known as ALS, or Lou Gehrig's disease), Alzheimer's disease, and Parkinson's disease.

5.3.1. Integrated Exposure Assessment

Integrated exposure assessments examine how the environment and pollutants affect the health of certain targets, such as human health (Behar, 1979). These assessments vary exposures from source to dose, over time and space, to multiple stressors, and from the molecular to ecosystem level in order to learn more about exposure effects in different settings. Such assessments aid with forecasting, preventing, and mitigating exposure that leads to adverse human-health or ecological outcomes by helping scientists create early warning systems and predictions. This helps to provide accurate information to resource managers, health departments, tourism operators, and the public regarding the potential for HAB-related illness and to prepare for controlling the HAB. This also helps groups prepare for potential economic impacts by giving them advanced time to plan budgets. Partnerships among NSF, the National Aeronautics and Space Administration (NASA), CDC, NOAA, EPA, and USGS, have resulted in various programs to assist managers in their immediate response to HAB events (Jewett et al., 2007; National Research Council, 2012). Basic and applied research and management strategies for HABs through these partnerships have advanced surveillance and predictive systems that "describe, reconstruct and forecast real world exposures more accurately and efficiently" (National Research Council, 2012). Integrating the scientific constructs of this effort brings existing strengths of HABHRCA-mandated science to scientific foundations that are now linking human and ecosystem health.

5.3.2. Drinking-Water Treatment

Removing toxins from drinking water is essential to protecting public health. Standard treatment procedures in drinking-water treatment facilities can remove or inactivate some algal toxins, but additional treatment technologies often are necessary, as standard processes may not be able to remove all algal toxins, especially from water sourced in the midst of a bloom (EPA, 2015b). Experts suggest that utilities and treatment facilities should develop individualized treatment plans for purifying drinking water, as toxin removal is dependent on toxin type and whether the toxin is dissolved in water or inside cyanobacterial cells (Westrick, 2008).

Cyanobacteria and their toxins have been found in drinking-water systems in the United States, but there have not been Federal guidelines, standards, or regulations concerning the management of HAB toxins until recently. In June 2015, EPA published Drinking Water Health Advisories for the cyanotoxins microcystins and cylindrospermopsin to assist Federal, state, and local officials, as well as managers of public or community water systems, in protecting public health. The advisories include information regarding available treatment techniques and analytical methods for these particular cyanotoxins. Furthermore, the Safe Drinking Water Act, as amended in 1996, requires the EPA to publish, every 5 years, a list of unregulated contaminants that are not subject to any proposed or promulgated national primary drinking-water regulations, are known or anticipated to occur in public water systems, and may require regulation. This list is known as the Contaminant Candidate List (CCL). The EPA's Office of Water included cyanobacteria and cyanotoxins on the first and second CCL (CCL 1, 1998; CCL 2, 2005). EPA also included on CCL 3 (CCL3, 2009) and the proposed CCL 4 (CCL4, 2015) anatoxin-a, cylindrospermopsin, and microcystin-LR for consideration.

Recent events, such as the massive algal bloom that developed on the Ohio River in early fall of 2015, have demonstrated the challenge that treatment facilities face in providing safe drinking water when encountering extreme HAB events. Over the last 2 years, the impact of HAB-related microcystins on several drinking-water-treatment facilities supplied by source water originating from Lake Erie were measured by EPA's Office of Research and Development. Results showed that certain treatment techniques can break intact cyanobacterial cells, thereby releasing toxins into drinking water during the treatment process. More positively, results also showed that, if operated properly, conventional water treatments designed to reduce turbidity effectively reduce intact algal cells.

Federal and state collaborations lead to re-opening of one of world's largest shellfish beds

Georges Bank, one of the largest clam beds in the world, was closed indefinitely to the harvest of shellfish in 1990 due to the risk of PSP toxins negatively impacting the shellfish industry. The situation became worse when an additional 15,000 square miles of Federal waters were closed in 2005 due to the presence of toxins. At this time, the clam industry reached out to FDA, NOAA, the Commonwealth of Massachusetts, academic institutions, and toxin test-kit manufacturers to seek a solution that would permit access to valuable shellfish resources. The resulting outcomes included the development of a test kit that could be used at sea by fishermen to determine when and where clams were safe to harvest, and a second test of the harvested shellfish performed by a laboratory. The onboard testing serves as an economic protection for the fishermen, while the second test is conducted by a National Shellfish Sanitation Program-approved laboratory to provide confirmation that the product harvested is safe for consumption. This collaborative effort resulted in the reopening of a large portion of Georges Bank to the harvest of surf clams and ocean quahogs in 2013, allowing the harvesting of billions of dollars' worth of clams, with control strategies implemented to ensure seafood safety.

While EPA regulates drinking water (i.e., tap water), FDA is responsible for regulating bottled drinking water. There are no specific regulations on cyanobacteria or their toxins in bottled water; FDA, however, requires bottled-water producers to process, bottle, and hold bottled water under sanitary conditions; protect water sources from bacteria, chemicals, and other contaminants; use quality-control processes to ensure the bacteriological and chemical safety of water; and test source water and products to ensure that they are free of contaminants.

5.3.3. Toxin Detection Methods

HAB research, management, and impact mitigation depend on accurate and timely detection of harmful HAB toxins. Robust, quantitative methods for identifying and measuring toxins are critical to assuring the safety of seafood, drinking water, and recreational areas, as well as to supporting clinical testing and confirming wildlife and domestic-animal poisonings. The complexity of algal toxins increases with the occurrence of multiple classes of HAB toxins and a diversity of active metabolites in ecosystems. Furthermore, algal species toxicity can vary from undetectable to high levels depending on strain composition and physiological status. This complexity makes it difficult to fully characterize the nature and extent of HAB impacts, and to develop effective and targeted methods to ensure the safety of food, water supplies, and recreational areas.

A variety of analytical methods are available for characterizing HAB toxins, some of which are designed to identify multiple toxin classes and other factors in a single sample. Identifying the species responsible for a given HAB can be helpful in narrowing the search for the possible toxin(s) that the HAB could be producing. Collaborations and common quality assurance and quality control protocols in lab and field work and strategy across Federal agencies and external organizations are essential in developing viable, rapid, and cost-effective toxin-detection methods.

5.4. Monitoring

Monitoring reduces risk, mitigates impacts, and providesaccurate information and predictive data for making management decisions related to HABs and hypoxia. Monitoring data support multiple goals, from tracking the sources of pollutants, to regulatory action, to predictive modeling and forecasting. Long-term nutrient monitoring is performed by numerous Federal and state agencies to track nutrient amounts and trends in discharge. Federal programs such as those carried out through USDA, EPA, and USGS focus monitoring programs on edge-of-field areas, as well as streams and rivers, in order to determine nutrient source, transport, and ultimate fate, which in turn enables program managers to track progress towards achieving the nutrient-reduction goals needed to lessen hypoxia. Routine monitoring programs for algal blooms and for the presence of HAB-derived toxins in shellfish were established under the National Shellfish Sanitation Program and conducted by many states, with as-needed assistance from the FDA.

Event-Response Programs

Event-response programs provide an early action to assist managers in their immediate response to HAB events in order to protect human and environmental health. Several Federal agencies share event-response capabilities.

- *The National Park Service (NPS) is developing a website for use by national parks on how to respond to HAB and hypoxia events. This website will be used by all park natural-resource managers to report events in a park or area near a park. Additionally, the agency is developing protocols for all parks for reporting any effects on wildlife, including but not limited to marine and terrestrial mammals, fish, birds, and shellfish. The NPS is also developing an event database to track the HAB and hypoxia events that occur in national parks*
- *NOAA's Marine Mammal Health and Stranding Response Program and the Working Group on Unusual Marine Mammal Mortality Events (NOAA National Marine Fisheries Service; U.S. Fish and Wildlife Service; Marine Mammal Commission, and EPA) investigate unusual mortality events.*
- *The USGS National Wildlife Health Center provides sample handling and project coordination for investigating wildlife disease or mortality events.*
- *FDA assists states with sample collection and analysis when marine biotoxins are suspected in state waters, and is the primary responder to blooms in Federal waters.*

- *The HABHRCA-mandated NOAA HAB Event Response Program provides funding for state managers and researchers investigating HAB events.*
- *The NOAA Analytical Response Team provides a formal framework through which coastal managers may request immediate coordinated assistance during HABs with species identification and toxin analysis.*
- *CDC has funded up to 10 states to develop programs to respond to HAB-related public health issues.*

Additional Federal programs and activities support the development of regional HAB- and hypoxiamonitoring programs, further assisting states. For instance, the EPA, states, tribes, and other partners are conducting a series of surveys of the Nation's aquatic resources. Often referred to as "probability-based surveys," these studies provide nationally consistent and scientifically defensible assessments of our Nation's waters by water-body type (e.g., rivers and streams, lakes, wetlands, or coastal waters), and can be used to track changes in condition over time. Each survey uses standardized field and lab methods and is designed to yield unbiased estimates of the condition of the whole water body being studied at a national scale and across broad, ecologically similar regions. Likewise, the survey can be used to help managers adapt how they manage resources in the short- and long-terms. This helps to reduce risk and helps to mitigate impacts, should any occur.

Phytoplankton monitoring is increasingly being used by state agencies to provide warnings and precautionary closures of harvest areas before shellfish become toxic. More specifically, in 2010, Wisconsin used modeling data to establish numeric water-quality standards for phosphorus in rivers, streams, lakes, reservoirs, and state portions of the Great Lakes (*Mississippi River/Gulf of Mexico Watershed Nutrient Task Force 2015 Report to Congress*). Likewise, new remote-sensing algorithms developed by NIEHS-supported scientists provide improved prediction, detection, and mapping of Florida red-tide events.

Responsibility for monitoring dissolved oxygen in most coastal, freshwater, and Great Lakes bodies of water lies with states or with Federal-state partnerships that utilize state monitoring programs to track water quality. For many of these areas where hypoxia occurs, fairly rigorous methods have been implemented for measuring dissolved oxygen and conveying this information to scientists and the public. The two largest hypoxic zones in the United States, located at the mouth of the Mississippi River and on the Oregon continental shelf, however, occur beyond the limits of state waters and, thus, rely on Federal support for monitoring. Monitoring informs coastal managers about water-quality conditions in the ecosystems they oversee, and it helps support the development and verification of ecosystem simulation models used to guide management decisions (CENR, 2010).

Nutrient loads delivered to a water body are calculated from nutrient concentrations and freshwater inflow. Nutrient concentrations in the water are generally monitored on monthly and on long-term bases, showing how nutrient concentrations vary over time. This helps managers and policymakers target specific times of year for developing effective mitigation strategies. This sampling schedule, however, can miss important events such as periods of high rainfall, when increased runoff brings more nutrients and can cause wastewater treatment plants to overflow, delivering greater-than-normal loads of nutrients to the water body. This sampling methodology can also miss pollution that comes directly from the air or groundwater. Hence, to better understand the small-scale changes in nutrient load and concentrations that can contribute to hypoxia, researchers have been investing in continuous nutrient measurement. For example, the USGS has 104 sensors nationwide (with over 40 sensors in the Mississippi River Basin alone) that measure nutrients on an hourly basis to more accurately determine nutrient loads to coastal bodies of water. This sensor network is operated through partnerships with other Federal and state agencies.

5.4.1. Observing Systems

Observing systems provide diverse streams of physical, chemical, and biological data for predicting HAB and hypoxia events in aquatic ecosystems in our Nation's oceans, coasts, freshwater bodies, and Great Lakes. Deployment of multiple observation platforms and sensors is required for monitoring the spatial and temporal resolution of phytoplankton that cause HABs and hypoxia. Satellites, moored ocean platforms or buoys, autonomous vehicles, and human observer networks provide data that can be assimilated into numerical models to develop forecasts and seasonal predictions of HABs and hypoxia events.

Having these technologies readily available helps communities prepare for HABs and hypoxia events and mitigate the effects of these phenomena on human and animal health, as well as on the economy. For instance, some species of *Pseudo-nitzschia* create a potent neurotoxin, domoic acid, which accumulates in filter-feeding fish and shellfish. When ingested, usually through seafood consumption, domoic acid can cause ASP, a severe and sometimes fatal illness. There are sensors for *Pseudonitzschia* and domoic acid that can be deployed on Environmental Sample Processors (ESPs) to detect the species and toxin as a bloom develops. Most recently, this was useful during an unprecedented, wide-spread *Pseudo-nitzschia* bloom along the west coast of the United States, providing communities with early warnings about potential human- and animal-health risks, and allowing managers to close proactively fisheries to harvest.

5.4.1.1. Eyes in the Sky

The physical and visual properties of water can reveal important insights about the biology of a water mass, as well as its potential to support HAB development and movement. A variety of airborne (aircraft) and spaceborne (satellite) commercial and government platforms are currently available for conducting remote water-quality evaluations (Reif, 2011). Airborne sensors measure the light and heat emitted from upwelling regions over large geographical areas in narrow bands. They provide data from visible and near-infrared wavelengths. Spaceborne sensors allow for consistent and frequent measurement of the light and heat from the Earth over moderate spatial resolution on regional and global geographic scales. Researchers have used data generated by such sensors effectively to detect phytoplankton-bloom activity and to inform models forecasting the development and movement of blooms on regional scales. The EPA is developing the Cyanobacteria Assessment Network (CyAN) mobile application using satellite-derived water-quality information—such as cyanobacteria cell-count concentrations for a local lake—to help water-quality managers make assessments (Wynn et al., 2013; Lunetta et al., 2015).

5.4.1.2. Eyes in the Water

There is a critical need for real-time, in-water observations of physicochemical properties (including dissolved oxygen), as well as of HAB species and toxins, to provide data streams for assimilation by predictive models. In-water sensors can take continuous measurements, supporting detection of physical and chemical indicators of HABs and hypoxia events. In-water observation systems—including moored observation platforms and autonomous underwater vehicles, or "gliders"—are well-developed and fully integrated into observing-system infrastructure in many areas. For example, in the Great Lakes, real-time information collected by a pilot observing-system buoy located in Lake Erie near Cleveland, Ohio, is provided to local drinking-water-processing managers. This information gives hourly updates of decreasing dissolved oxygen, internal wave status, and temperature profiles, which are used to determine the depth of the temperature gradient (thermocline) and dissolved oxygen levels. The data may warn of potential hypoxic events from July through October and give managers time to prepare alternate processing methods.

Although timely reporting of conditions that may lead to HAB or hypoxia events is indeed useful, real-time, autonomous measurements of algal cell and toxin concentrations and of dissolved oxygen can provide the level of detail required for accurate early warnings as well as for the current status of an ongoing HAB or hypoxia event. Technologies with such capabilities are now commercially available and are being deployed as experimental components of several observing networks to detect HAB or hypoxia development or assess the toxicity of an event. Over the next 5-10 years, maturation of observation-system technologies will facilitate their transition to operational status and full integration within regional observing systems.

5.4.1.3. Eyes on the Coasts

Citizen engagement plays an important role in HAB and hypoxia surveillance, and therefore in predicting and mitigating HABs and hypoxia events. A number of relevant citizen-science programs are run across the country. NOAA's PMN program links volunteers who monitor for marine phytoplankton and HABs with professional scientists, effectively building a more-informed public while expanding the reach and resolution of HAB monitoring. Over 200 PMN volunteers sample more than 140 sites in 17 states and the U.S. Virgin Islands. Volunteers are trained by NOAA staff on sampling techniques and identification methods for marine and freshwater phytoplankton. Since the program began in 2001, volunteers have reported more than 120 algal blooms and 7 toxic events.

NOAA has worked with other Federal agencies to improve the PMN program. The PMN program prior to 2010 focused on marine phytoplankton species; however, beginning that year, CDC and NOAA agreed to collaborate on a study to examine if PMN methods could be applied to monitoring HABs in freshwater lakes. The pilot study included 12 lakes in Minnesota and Wisconsin, with 34 sample locations identified by CDC. Additionally, CDC provided funding to adapt the existing marine methodology, and to provide training and equipment for 39 volunteers. The study targeted four freshwater cyanobacterial HABs and two nontoxic bloom species, and researchers analyzed 275 samples between June and October 2010. Of note, 97 percent of the samples were found positive for at least one target species. Moreover, all

Wisconsin lakes contained presence of the toxic species Microcystis during the sampling season. Three major blooms were confirmed to be composed of toxic species, and several "green water" blooms that were presumed to be toxic were composed of nontoxic species. This pilot study demonstrated the effectiveness of PMN method for volunteer monitoring of toxic versus nontoxic freshwater HABs. It also provided the foundation for the NOAA and EPA to respond quickly to establish freshwater PMN sites in Lake Erie in 2015 and to expand to Lakes Superior, Michigan, Huron, and Grand Lake St. Mary in 2016. NOAA entered into an Interagency Agreement with EPA to expand the PMN to freshwater HABs in the Great Lakes region beginning in 2015 (Appendix 3). Other examples of long-term monitoring programs include the Olympic Regional HAB (ORHAB) and SoundToxins partnerships, which provide early warnings of HABs on the outer Washington coast and Puget Sound, respectively.

5.4.2. Models and Forecasts

New and expanded observing capabilities have grown the potential for early detection and characterization of HABs and hypoxia events. Yet given the considerable temporal and spatial scale of HABs and hypoxia events, direct observations alone cannot provide sufficient warning and information about all HABs and hypoxia events in all areas in which such events may occur. Accordingly, a range of models exist to complement the early warnings and forecasts derived from direct observations (McGillicuddy, 2010).

For instance, models have been used to study the sources of nutrient pollution—a key contributing factor to both HABs and hypoxia events. Models such as the USGS' SPAtially Referenced Regressions on Watershed (SPARROW), or USDA's Agricultural Policy Environmental Extender (APEX) and Soil and Water Assessment Tool (SWAT) (used in CEAP), provide consistent approaches to estimating nutrient sources in coastal areas. Long-term data on nutrient enrichment has been used in other SPARROW models, which examine nutrient sources for freshwater streams, lakes, and coastal areas, and in new methods for detecting trends in nutrient loads.

These models indicate (Figure 5.1) that agricultural input (i.e., manure, fertilizer, and legume crops) is the largest nitrogen source for the conterminous United States (50 percent of the total), atmospheric deposition contributes 27 percent, and urban sources contribute about 22 percent (7 percent from urban areas and 15 percent from wastewater-treatment plants). Anthropogenic phosphorus inputs are proportionately lower because of larger contributions from background sources, including weathering of phosphorus-rich soils and bedrock, as well as erosion of phosphorus-rich stream sediments previously deposited from upstream sources. The relative contributions of sources vary across and within watersheds. For example, while agricultural, atmospheric, and urban sources are the predominant nitrogen sources in the North/Mid-Atlantic, Great Lakes, and Pacific Northwest, respectively, approximately 50 percent of the annual nitrate load into the Chesapeake Bay enters the watershed through streams via groundwater (Sanford, 2013).

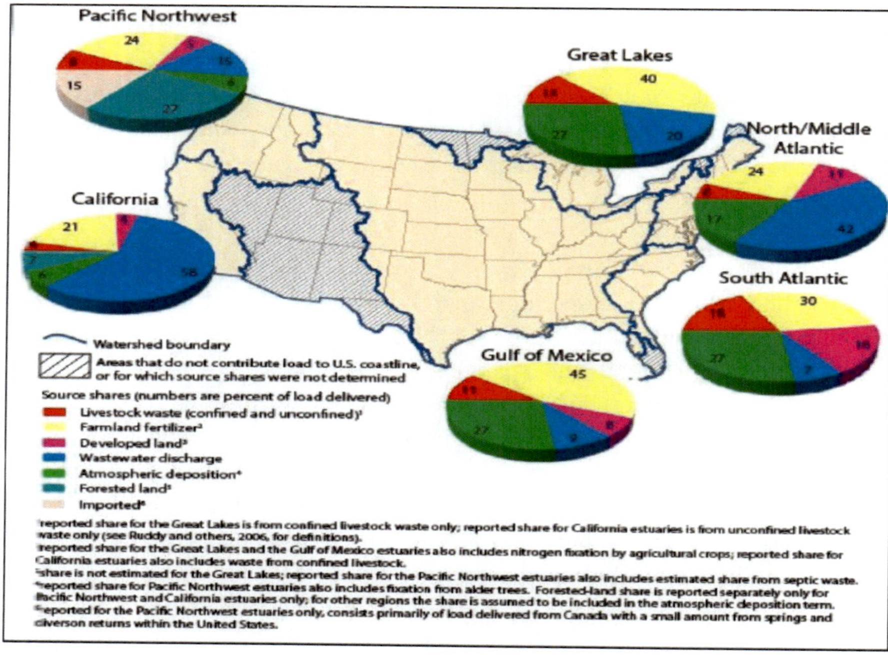

Figure 5.1. Source shares of nitrogen load delivered to the coastline of the conterminous United States, by region. Maps of nitrogen and phosphorus yields, loads, and watershed rankings with nutrient source information for a state or large river basin can be accessed using the USGS SPARROW model. SPARROW results for key estuaries have also been tabulated, and the SPARROW Decision Support System can simulate nutrient reduction scenarios for selected watersheds.

Numerical models provide the foundation of NOAA's HAB forecasting system developed for *Karenia brevis* blooms in the Gulf of Mexico (Stumpf et al., 2009). HAB forecasts are made twice weekly during bloom events, using a combination of satellite-derived image products, wind predictions, and a rule-based model derived from previous observations and research. Blooms are detected and defined using ocean color satellite images. Bloom transport is then predicted using hydrographic modeling with particle-tracking techniques.

Seasonal forecasts for the Gulf of Maine and the Chesapeake Bay show the size, movement, and potential severity of specific types of HABs and hypoxia events in those regions (Allen et al., 2015). The models are based on a number of factors, including water motion driven by winds, tides, stratification, river runoff, and upwelling. They also take into account temperature, salinity, sunlight, and nutrients. Seasonal forecasts in the Chesapeake Bay specifically draw on the relationship between reducing nutrient loads and a resulting decrease in late-summer hypoxia. In the long-term, analysis of 60 years of monitoring data showed that reducing nutrient runoff into the Chesapeake Bay has reduced hypoxia in the late summer, but large-scale climatic forces are leading to early-summer hypoxia formation. Despite the influence of climatic factors on hypoxia, nutrient management is a key mitigation strategy for improving the water quality of the Chesapeake Bay (Kemp et al., 2005). In the Chesapeake, the Gulf of Maine, and other locations, improved information and forecasts of same-season variability allow farmers and other agricultural producers to better manage fertilization, tillage, drainage-water management, and other systems. This information is also useful for water managers and other industries to use for advance planning.

The expansion of NOAA's HAB- and hypoxia-forecasting systems—including systems that provide advanced ecological forecasts, early-warning bulletins, and other early-warning products—has had a significant impact upon the ability of communities and regions to prepare for these events, and represents one of the singular accomplishments since the last authorization of HABHRCA. A list of these forecasts follows:

HAB Forecasts:

- *West Florida Shelf (operational)*
- *Texas (operational)*
- *Lake Erie (in transition to operations – 2016-2017)*
- *Gulf of Maine (demonstrational/experimental)*
- *Puget Sound (demonstrational)*
- *Washington Coast (demonstrational)*
- *California (demonstrational)*
- *Chesapeake Bay (future)*

Hypoxia Forecasts:

- *Gulf of Mexico (demonstrational/experimental)*
- *Chesapeake Bay (demonstrational/experimental)*
- *Puget Sound (demonstrational/experimental)*
- *Green Bay (demonstrational/experimental)*
- *Narragansett Bay (demonstrational/experimental)*

5.5. Public Advisories

Providing advance notice of the possibility of HABs or hypoxic conditions is a critical step in addressing human-health concerns and helping to mitigate the impacts of these events. Reaching out to communities and making readily accessible and available information about HABs or hypoxia events enhances consumers' awareness of potential hazards of these phenomena and enables managers to make science-based decisions, thereby helping to protect the health of communities. Advisories also allow tourism operators and other business owners who may be affected by HABs and hypoxia to prepare accordingly.

6. Challenges in HAB and Hypoxia Management

There are many gaps in the research and management communities' knowledge of HABs and hypoxia events. The complex causes and effects of these events understandably present many challenges to researchers, managers, and decision-makers in HAB and hypoxia prevention, control, and mitigation.

The ability to predict the response of HABs and hypoxia to changes in nutrient loading and other drivers, such as climate, is difficult but critical. Land-use change, changes in precipitation patterns, and altered hydrology collectively have led to increases in stream discharge, as well as baseflow (Zhang and Schilling, 2006; Schottler et al., 2014; Scavia et al., 2014). The management of nonpoint nutrient pollution, in particular, requires a better understanding of nutrient discharge to surface water and groundwater. Understanding the delivery modes of nutrients to the coasts would allow policymakers to better characterize the lag time between nutrient reduction on land and improvements in the quality of in surfacewater and groundwater.

Similarly, improving understanding of the impact of legacy phosphorous and sediment within river systems or lakes is significant for predicting and addressing HABs and hypoxia events. Researchers need to learn more about how the cycling of legacy phosphorous or sediment is affected by reductions of new sediment or phosphorous entering the system, and to establish the causes of annual nitrate-level fluctuations. Scientists also need better information on contributing nutrient sources such as fertilizers, urban inputs, and wastewater to develop nutrient control and monitoring strategies.

Predicting species composition of algal blooms is challenging, as is determining the circumstances in which a potentially toxic bloom will produce toxins. Improving knowledge of the response of HABs and hypoxia to nutrient-reduction strategies and changes in weather patterns and climate will help to develop decision support tools that are grounded in science.

6.1. Environmental, Economic, and Social Challenges

The environmental, economic, and social impacts of HABs and hypoxia in marine and freshwater regions are incompletely understood. Research is needed to identify and quantify how HABs and hypoxia affect ecosystems, such as through habitat loss and food-web disruption, so that cost-effective and scientifically-sound management strategies—including strategies for reducing nutrient inputs—can be developed and implemented to achieve healthy water quality in rivers and streams, lakes and reservoirs, and estuaries and near-coastal marine waters. There is also limited research examining the short- and long-term effects of HABs and hypoxia on communities. The lack of clarity on the impact of HABs and hypoxia represents the largest source of scientific uncertainty for policy development in most systems. This is a compelling driver for action, as the lack of information of costs and benefits of various management options presents a significant stumbling block to implementing mitigation plans in many regions.

6.2. Monitoring Challenges

The challenges of monitoring include sustaining monitoring programs, ensuring consistency of monitoring methods, implementing rigorous benchmarks of quality assurance and quality control, and synthesizing and making available monitoring data.

The successful deployment of nutrient sensors for continuous, real-time monitoring depends on ongoing evaluations of sensors and technologies that communicate data, development of nutrient-monitoring standards and protocols, improvements in tools for data-quality assurance and data visualization, and personnel training.

Toxin monitoring could be advanced by having sensors that can directly detect toxins *in situ* and in real time. This would help defend against human or animal exposures or effects in marine, fresh, and recreational waters, and at drinking-water sources. Even though there has been some success in detecting toxins with *in situ* sensors, only a few toxins or derivatives are well understood and categorized. Another toxin-monitoring challenge relates to knowledge of the varieties of cyanotoxins and derivatives to understand which are most toxic. Developing new or improved sensors that are able to differentiate different types of HABs, and even toxic vs. non-toxic groups of HABs, would be beneficial.

Challenges in integrating data on HABs and hypoxia are due to the variety of stakeholders collecting and using this information. While increased collaboration helps to develop new methods of data collection and analysis, it also increases margins for error and slows down systems due to additional layers of process and review.

6.3. Research and Modeling Challenges

Researchers are trying to address the conditions that contribute to, cause, or end HABs, HAB toxins, and hypoxia with improved models and more collaborative, interdisciplinary research. Given the complexities of the number of HAB species, and the evolution of how drivers interact to lead to HABs and hypoxia, many challenges remain.

Specific studies in functional dynamics at the microbial levels are needed because algal cell processes may hold the key to understanding the production of HAB toxins. Conditions in an ecosystem can affect the ability of a cell to reproduce or to exist long enough to form a bloom. Additionally, certain species of algae feed upon others, or simply are stronger competitors for resources, allowing these species to proliferate and form blooms (Marinho et al., 2013; Chakraborty and Feudel, 2014). Scientists at this point are unclear as to how exactly the specific dynamics function at microbial levels, however.

A major challenge in modeling HABs and hypoxia is the complex nature of how nutrients and other factors influence HAB and hypoxia development.

While the processes that produce these events are understood generally, local factors strongly influence the severity and other characteristics of HABs and hypoxia events. Wind, rain, water-column stability, buoyancy changes in cells, underwater irradiance, and lake morphology all impact the ability of models to predict blooms. Furthermore, making predictions is further compounded as these factors must be modeled in conjunction with nutrient levels and the varying, complex conditions of a given body of water.

Factors that affect local nutrient conditions include nutrient inputs from watershed, atmospheric, and other sources; nutrient processing and recycling on local and broader scales, once these nutrients enter a water body; and the relationship between nutrients and algal productivity. Improved water quality and ecosystem models are needed to better assess and predict effects of nutrient management on downstream water quality, and the impacts of hypoxia on fish populations and food-web dynamics, in order to refine management strategies for protecting coastal economies and aquatic resources.

Models can estimate external nutrient inputs and simulate nutrient processing and algal production, but every model relies on monitoring data, which can be variable in quality. Every model undergoes different types and levels of quality control. Additionally, few models are available that link HABs and hypoxia events with impacts on social well-being, economics, and human health, and therefore there is little actionable information for use in developing comprehensive management plans.

Models must take into account at least three components: nutrient sources from the land, atmosphere and ocean; how nutrients are processed on local- and estuary-wide scales once they enter a water body; and how the nutrients influence the growth of algae species. Presently, there are models that estimate the loads of nutrients that enter a water body from land, air, and sea. Likewise, there are models that simulate the way that nutrients move around and through a water body once they enter it, and still others that account for algal growth and changes in nutrient concentrations as the algae use the nutrients. Many of these models are highly complex and do a good job of representing nutrient movement from land to water, and, separately, nutrient movement in the water. There are limited models or modeling systems that integrate the two components of nutrient movement and provide information on the associated impacts on social well-being and economic and human health that can inform comprehensive management plans. Overall, expansion of river, stream, and watershed monitoring, coupled with increased modeling information, can advance understanding of nutrient sources, transport, and ultimate destination, thereby enabling better tracking of progress towards achieving nutrient-reduction goals.

6.4. Human Health Impacts Challenges

To reduce human-health impacts caused by exposure to HABs, scientists must understand the causes of HABs and the level of awareness of associated human-health risks. Specifically, researchers must study how different toxin concentrations and exposure scenarios— including exposure type (inhalation, dermal, or ingestion) and duration (acute or chronic)— affect humans, as well as the concentrations of different toxins needed to trigger adverse health effects. Identifying individuals who are at higher risk for experiencing adverse health events related to harmful algae exposure also is a priority.

The WHO has developed guidelines for lifetime drinking-water exposure to microcystin-LR, and for acute exposures to cyanobacterial cells in recreational waters. Additionally, three cyanotoxins (microcystin-LR, cylindrospermopsin, and anatoxin-a) have been proposed for EPA's CCL for evaluating public health risks in drinking-water systems. In June 2015, the EPA released 10-day Health Advisories (HAs) for two cyanobacterial toxins: microcystins and cylindrospermopsin. The advisories describe concentrations of the two algal toxins in drinking water at or below which adverse health effects are not anticipated to occur over a 10-day exposure period. The EPA also released a support document containing recommendations for managing cyanotoxins in drinking water, which includes a framework for cyanotoxin risk-management efforts. Some states and other countries have developed their own guidelines for certain toxins (Graham et al., 2009; Chorus, 2012). The basis for these guidelines and recommendations rests primarily upon knowledge of acute and subacute toxicity of cyanotoxins; there are still large knowledge gaps regarding the effect of chronic exposure to cyanotoxins on human and animal health.

Human and animal exposure to HABs can involve multiple toxins (either simultaneously or sequentially), making it difficult to track and test for toxin source. Exposure to multiple toxins can result in both synergistic and additive harmful effects: one toxin might target the liver, which provides detoxification for a second toxin that targets a different organ system, or two toxins might target the same organ system by different mechanisms. Studies on the effects of multiple toxins require considerable resources and manpower, but they can be important, given the possibility of such effects.

Impacts of HABs on wildlife in managed and natural environments present a particular challenge that needs to be addressed. While drinking water intended for human consumption can be treated to remove toxins, animals consume and come into contact with untreated aquatic waters and therefore are at higher risk of being directly exposed to HABs. Furthermore, HAB-related animal illnesses may be underreported (Backer et al., 2013) because clear and easy-to-use reporting tools and communication mechanisms are not readily available to pet owners, veterinarians, wildlife and fisheries personnel, medical and public-health officials, and others. Addressing this issue would also lead to additional human benefits. Hence animal deaths and illnesses resulting from HABs or hypoxia events can serve as sentinels of potential human-health risk and, if such incidents are reported quickly, can trigger appropriate responsive action that mitigates risk (van der Schalie et al., 1999).

7. Research Plan and Action Strategy: Recommendations

Progress to this point has demonstrated that science-based decision-making can effectively help prevent and mitigate the effects of HABs and hypoxia events. Keeping up with the evolution of HAB and hypoxia events and the continual discovery of new HAB species, however, is challenging. While Federal agencies have successfully implemented and planned many programs related to HABs and hypoxia, there are many ongoing gaps in policies, programs, and science that require attention. The IWG-HABHRCA recommends the following Federal research plan and action strategy. An implementation plan is pending.

1. Adding to and Improving Scientific Understanding

a. Develop certified reference materials (CRMs) and other standardized and validated detection and analysis methods for HAB toxins.

HABHRCA calls for increasing the availability of analytical facilities and reference materials, and improved ability to predict the onset of toxicity. The biggest challenges to analyzing HAB toxins include the limited availability of CRMs and standardized and validated analysis methods,[3] all of which are essential to ensuring consistency and comparability of results between laboratories and other testing environments. CRMs are lacking for many of the classes of toxins that impact American waters, as well as those of international origin that may enter the United States' food supply through seafood.

While progress has been made herein, the scope of the remaining need is greater than any one agency. The IWG-HABHRCA recommends establishing an interagency team to examine current capabilities, and to establish a strategy for developing intra- and interagency methods and approaches for developing and sustaining CRM availability. An outcome of this IWG could include establishing a network of several reference laboratories.

Additionally, there is a need for commonly available, affordable methods for toxin analyses—quality-assured with CRMs—that can be implemented by a variety of user groups, and validated by analytical methods and inter-laboratory trials. These will ensure comparable results from the earliest to the final tiers of testing, thereby helping to improve the ability of all user groups to prevent, control, and mitigate HABs.

b. Conduct studies on toxins in food and on toxin mixtures.

While reported algal toxin-related human illnesses are often associated with the consumption of seafood, based on dietary history and symptoms, confirmation of foodborne illness remains a challenge. For instance, ciguatera fish poisoning is the most common HAB toxin-related seafood poisoning, yet a confirmed diagnosis is difficult due to a lack of accurate clinical methods for detecting ciguatoxins in biological specimens.

Likewise, human and animal exposures to HABs often involve multiple classes and metabolites of toxins, and there is a critical need for finding a better way to assess the relationships between these toxins. Although toxicological evaluations of toxin mixtures require considerable resources and manpower, such evaluations are essential given the real potential of additive or synergistic effects from multiple toxins. This precautionary approach ultimately will help managers and the government to issue advisories and make accurate policy decisions in the future.

c. Develop more-effective HAB suppression and control methods that have minimal environmental effects and lower cost.

Potential control techniques are needed to investigate further include increasing flushing rates, ultrasound, electrocoagulation, new and existing coagulants, and new algicidal or algistatic compounds. These techniques are used to break up HABs.

d. Understand the influence of climate change, atmospheric deposition of nutrients, and other contributing factors on the occurrence, frequency, and severity of HABs and hypoxia.

Potential effects of climate change include rising average water temperatures and increased precipitation, which can support the development of HABs and hypoxia. Other impacts of climate change, including ocean acidification, interact with nutrient loading and eutrophication in the coastal zone, which may result in additional impacts to aquatic life (including aquaculture). The exact nature of these interactions and their impacts generally are not well understood and should be the subject of further research.

Scientists also need a better understanding of how other factors affect HABs and hypoxia. Particularly, there are knowledge gaps related to the effects of point and nonpoint sources of nutrients, such as atmospheric deposition of nitrogen or sediment and nutrient runoff, may lead to HABs and hypoxia events.

e. Develop case definitions for the spectrum of HAB-related illnesses and produce clinical therapeutic guidance for the spectrum of illnesses associated with exposure to HAB cells and toxins.

2. Monitoring

a. Strengthen long-term HAB and hypoxia monitoring activities.

In many cases, comprehensive monitoring programs are needed to provide enough coverage of both watershed and coastal waters to support scenario-based models used to determine progress of management actions and inform management and mitigation strategies. Comprehensive monitoring programs are also important for determining the effects of climate change on HABs and hypoxia, such as the potential impacts of increased ocean acidification on cellular toxicity. More comprehensive and coordinated monitoring programs are needed to better understand the relative contributions of different nutrient sources to HABs and hypoxia, which nutrient forms are most problematic, and the relationship of nutrients to the development of HABs or hypoxia and HAB toxicity. Expanded monitoring can help advance understanding of the relationships of nutrients with other factors that exacerbate HABs and hypoxia, such as increased water temperature.

In particular, there is a need to strengthen long-term monitoring activities conducted by USDA, USGS, NOAA, and EPA in the Gulf of Mexico watershed and coastal zone through implementation of the National Monitoring Network. In the dead zone of the northern Gulf of Mexico, lack of operational support has limited the execution and management capacity of an implementation plan for a comprehensive monitoring program designed to help scientists understand hypoxia causes and impacts. Furthermore, in the Great Lakes and other inland lakes, better monitoring is needed to help determine the relative impact of quagga mussels on in-lake responses to phosphorus-management strategies, and determine how sediments affect internal phosphorus cycling and loading and associated effects on HABs and hypoxia. A greater understanding of legacy phosphorus and sediment in Lake Erie in particular is needed to inform response strategies to more effectively address HABs and hypoxia events.

b. Integrate new monitoring technologies into emerging U.S. and global ocean-observation systems.

For a more complete understanding of HABs and hypoxia, new sensor technologies must be incorporated into observations systems to enable the collection of wide streams of environmental data that can be used to trigger early warnings and improve HAB or hypoxia characterization. Likewise, the development and deployment of more portable instruments by field observer networks will enable faster detection of algal cells and toxins to trigger early protective actions, and increase the spatial and temporal resolution of data on HABs or hypoxia events.

c. Develop a rapid-response strategy for assessing HAB exposure.

There is a need to establish a rapid sample collection and response protocol for detecting HAB toxins in humans and animals. Given how quickly many toxins are metabolized, protocols like this will strengthen diagnoses and will help to establish the best treatment methods.

3. Prediction

a. Develop, improve, and validate HAB and hypoxia models and remote sensing.

Predictive models are critical for understanding HAB and hypoxia effects on ecosystems, and for enabling effective forecasting of and response to toxins in drinking and recreational waters. Improvement and expansion of HAB and hypoxia modeling efforts and remote satellite sensing can also help to improve early-warning technologies to minimize human and animal exposure to HAB toxins, and to assess progress of management actions and mitigation strategies. Researchers also need models to help elucidate the fate and effects of HAB toxins, as well as the long-term risks to ecosystems from exposure to HAB toxins. Furthermore, the ecosystem impacts of hypoxia on fisheries can differ by fish species, requiring both population-level and ecosystem-level modeling approaches to fully quantify ecological andr socioeconomic consequences. In addition to *in situ* monitoring via moored platforms and gliders, these efforts should include satellite and aircraft remote-sensing capabilities for HABs and hypoxia that can be utilized in monitoring and management programs.

b. Develop enhanced surveillance for human and animal exposure, illnesses, disease, and deaths resulting from HAB toxins.

This should build on existing surveillance systems, such as the HABs module in the National Outbreak Reporting System of CDC.

4. Emphasize Stakeholder Engagement and Socioeconomic and Understanding

a. Improve communication and coordination among health and environmental agencies so that reports of HAB-associated animal poisonings are used as an indicator of potential human-health risk. Develop science-based guidelines for cyanotoxins.

Researchers must develop science-based guidelines for cyanotoxins, to help managers make effective decisions and improve their public communications. Managers should work to develop communications and outreach programs to make information on HAB poisoning syndromes available to the medical and veterinary community in a timely manner.

Wildlife and fisheries personnel are usually the first responders to mass wildlife deaths associated with HABs. Among some of these occurrences, animal deaths and illnesses may be used as potential sentinels of human-health risk, as the animals have higher direct exposures to HABs in natural waters that may be used for food provision, recreation, or as a source of drinking water by humans. Communication among wildlife, veterinary, medical, and public-health officials needs to improve to increase HAB surveillance and to protect the public health.

b. Identify susceptible populations at higher risk for HAB-associated adverse health effects.

Identify people and animals at increased risk for HAB-related adverse health effects and target them for interventions. To prevent illnesses, researchers must understand their cause. People in areas of chronic exposure may be aware of the risk of HAB toxins but lack the means necessary to prevent illness, therefore leaving them at higher risk of HAB-related illness.

c. Expand stakeholder engagement.

The strategies to reduce human impacts from HABs and hypoxia involve a combination of improving the availability of existing resources, better coordination and communications among stakeholders, and setting priorities for programs to facilitate technology transfer. There is a critical need for rapid, cost-effective monitoring tools that can be used to involve a large group in monitoring HABs and hypoxia. Improvements in monitoring tools and data sharing are needed to bolster the capacity of stakeholder groups in HAB and hypoxia prediction, detection, and response, which in turn will make it possible to implement the necessary management actions and focus toxin testing where it is most needed. Expanded stakeholder engagement is also important for educating managers, planners, decision-makers, and the public about what monitoring data can tell them about the risks presented by HABs. Public service announcements about the effects of HABs and hypoxia, and how to avoid and prevent them, can be helpful in catalyzing increased public engagement and action. Engaging land owners and agricultural producers within affected basins and watersheds is critical to increasing awareness of the issues and developing a shared strategy to inform decision making and more successfully HAB and hypoxia issues.

d. Evaluate socioeconomic impacts of HABs and hypoxia, and the costs of mitigation.

There are many knowledge gaps related to the socioeconomic impacts of HABs and hypoxia, including in inland lakes. Assessment of the socioeconomic impacts of individual HAB or hypoxia events is needed in order to determine which types require the greatest attention and resources.

Also needed are models of the socioeconomic costs of HAB and hypoxia impacts (on food, drinking water, recreation, natural resources, as well as aesthetic impacts and lost ecosystem services), and the cost effectiveness of prevention, control, and mitigation strategies, including nutrient reductions, to support decision-makers and inform the prioritization of mitigation approaches and goals. Developing these models may require recruiting resource economists and other social scientists to help assess HABs and hypoxia events.

5. Policies and Synergies with Other Programs

a. Continue and expand relevant research, management, and policy collaborations.

Many of the research initiatives mentioned in this report have been made possible by collaborations among Federal agencies, as well as with other state and local entities, and with the public. Larger, more complete datasets need to be developed and shared between agencies and outside of the government, and used in developing improved forecasting and decision-making products. As a broader diversity of sectors are involved in these collaborations, people will be more educated about the changes in their environment, enabling the most effective HAB and hypoxia prevention, control, and mitigation effort possible.

b. Develop guidelines and tests for HAB toxins in drinking and recreational water, and improve toxin removal during water treatment.

Continued efforts to make drinking water safer from HAB toxins should include development of real-time monitoring systems for toxins and cell fragments during water treatment. Investigations of other methods for removing toxins during water treatment— including enhanced coagulation technology, filtration effectiveness, and disinfectant byproducts—are similarly important.

8. Integrated Assessment of Great Lakes HABs and Hypoxia

8.1. Introduction

The intent of this chapter is to serve as a separate integrated assessment on HABs and hypoxia in the Great Lakes. HABs have occurred in all of the Great Lakes, causing detrimental effects like hypoxia or producing pathogens that affect wildlife and humans. Specific to the Great Lakes, some HAB species can harbor pathogens that include avian botulism, which kills fish and birds, and waterborne human pathogens. Most HAB events in the Great Lakes are composed of cyanobacteria (Appendix 1), and green algae (e.g., *Cladophora*) predominantly form the HABs that create algal masses.

HAB events occur in the summer months in the western basin of Lake Erie, Saginaw Bay, and Green Bay as well as in multiple small embayments and tributaries, such as the Sandusky and Maumee Rivers, Muskegon Lake, and Little Bay de Noc. Since the late 1990s, blooms have also occurred on the southern shore of Lake Ontario, and more recently in the waters of western Lake Superior. HAB species can be found in the nearshore waters of each of the Great Lakes.

Hypoxic zones occur most frequently in the central basin of Lake Erie and in Green Bay (Burns et al., 2005; Hamidi et al., 2013), but also episodically in western Lake Erie and Saginaw Bay (Bridgeman et al., 2006; Stow and Hook, 2013). Though hypoxia occurs naturally in Lake Erie's central basin (Delorme 1982), human activities, including nutrient loading, have made it worse (Scavia et al., 2014).

8.2. Causes of HABs and Hypoxia

This section of the report covers the causes, consequences, current research programs, and gaps in research and management with regard to prevention, control, and mitigation of HABs and hypoxia events in the Great Lakes. Since the 2004 HABHRCA authorization, substantial progress has been made with Great Lakes HABs and hypoxia events through expanded and improved policies, research, and management programs.

Point-source nutrient inputs to the Great Lakes were reduced to a great extent in the 1970s, which substantially reduced regional HABs and hypoxia events. Since the late 1990s, however, HABs and hypoxia events in the Great Lakes have had a resurgence, most notably in Lake Erie. Contributing factors include increased non-point, agricultural inputs of dissolved reactive phosphorus and both inorganic and organic forms of nitrogen, combined with changes in precipitation timing and intensity, which can affect nutrient transport and increases in stream discharge (Chaffin et al., 2013; Michalak et al., 2013; Scavia et al., 2014). Aging sewer systems and an associated increase in combined sewer overflows could also be a contributing factor, although this is less certain.

HABs in the Great Lakes likely are the result of complex interactions among multiple factors rather than any one specific factor (Smith et al., 2015). Phosphorus and nitrogen directly promote excessive algal biomass that can be associated with higher HAB toxin levels in some aquatic species (Hudnell et al., 2008; Davis et al., 2010; Horst et al., 2014). The presence of invasive dreissenid mussels in the Great Lakes appear to promote some HAB species, but the connection is not always clear (Vanderploeg, 2001; Conroy et al., 2005; Bridgeman and Penamon, 2010). Herbicides and pesticides that are washed into lakes during application or times of high precipitation can also promote Great Lakes HAB species by killing their natural competitors (Peterson et al., 1997; Lürling and Roessink, 2006). Hypoxia is caused when organic matter, including settled algal biomass, is decomposed by bacteria that consume oxygen, and it can also promote HABs by increasing phosphorus release from sediments (Correll, 1998).

In the Great Lakes, warm, calm water caused by seasonal temperatures and wind patterns can promote the separation of oxygen rich and depleted waters. Climate change may exacerbate one or many HABs and hypoxia events (Paerl and Huisman, 2008; Davis et al., 2009; O'Neil et al., 2012; Paerl and Otten, 2013) by causing even warmer air and water temperatures, and by causing changes in precipitation patterns that increase stream discharge and promote phosphorus and nitrogen loading into nearshore waters (Paerl and Paul, 2012; Michalak et al., 2013; Scavia et al., 2014).

8.3. Impacts of HABs and Hypoxia

HABs and hypoxia impacts in the Great Lakes are similar to those in marine and other freshwater systems. For more details about the effects of

HABs and hypoxia in aquatic environments, see the "Threats Caused by HABs and Hypoxia" section of this report.

8.4. Relevant Legislation and Funding Initiatives

In 2010, President Obama announced the EPA-led, interagency Great Lakes Restoration Initiative (GLRI), which was designed to increase efforts between 2010-2014 to protect and restore Great Lakes natural resources, and included HAB and hypoxia projects. GLRI was extended from Fiscal Year 2015 through FY2019, through the GLRI Action Plan II, and allows for further collaborations between Federal, state, and local authorities. After the 2014 Lake Erie HAB, $12 million in GLRI funding was provided to Federal and state programs to minimize HABs and hypoxia in the western basin of Lake Erie. Specific projects supported by this funding include upgrading controlled drainage systems, funding best management practices (BMPs) at livestock facilities, and planting cover crops. It also provides funding for the Environmental Quality Incentives Program, a voluntary program through USDA NRCS that provides financial and technical assistance to agricultural producers to plan and implement conservation practices that improve soil, water, plant, animal, air, and related natural resources on agricultural land and non-industrial private forestland.

On March 26, 2015, EPA announced the award of 14 GLRI grants totaling over $17 million, to fund projects that will improve Great Lakes water quality by preventing phosphorus runoff and solid erosion that contribute to algal blooms, and by reducing suspended sediments in Great Lakes tributaries. These projects focus on high-priority watersheds and receiving waters with high potential or known risk for HABs and hypoxia. The high-priority areas are Fox River-Green Bay, Saginaw River-Saginaw Bay, and Maumee River-western Lake Erie. Nutrient-abatement projects in other high-concern areas were also funded.

In 2012, the United States and Canada amended the Great Lakes Water Quality Agreement (GLWQA). The agreement addresses HABs and hypoxia, among other issues, via lake-wide management of nutrient inputs for more biologically relevant nutrient reduction targets. Nutrient allocations will be made at country, state, and provincial levels for watersheds that are at highest risk for HAB and hypoxia development.

The Agricultural Act of 2014 (commonly known as the new "Farm Bill") was passed in February 2014, and retained and expanded many existing programs to reduce nutrient and sediment loads into the Great Lakes Basin. A new program authorized under the legislation is the Regional Conservation Partnership Program (RCPP). RCPP encourages partners to join with agricultural producers in efforts to increase the restoration and sustainable use of soil, water, wildlife, and related natural resources. In particular, the Secretary of Agriculture selected the Great Lakes Region as a Critical Conservation Area, making more conservation funding available to build on existing strong partnerships in the Great Lakes Region and to provide approaches and tools for agricultural producers to better manage nutrients and sediment on agricultural land.

8.5. Current Agency Actions and Successes: Prevention, Control, and Mitigation

In the Great Lakes, modeling and monitoring nutrient inputs in Lake Erie are examples of a *prevention* strategy to reduce the inputs which may increase HAB and hypoxia. *Control* efforts address the immediate causes of HABs and hypoxia. This includes strategies that directly kill HAB cells or destroy toxins, physically remove cells and toxins from the water column, or limit cell growth and proliferation. *Mitigation*—which refers to responding to an existing or ongoing bloom by attempting to restrict, inhibit, or prevent associated undesirable impacts on the environment, human health, or human economies and communities—is the area of HAB management where the most immediate potential exists to reduce impacts, given that many such activities are already underway (Seltenrich, 2014). Improved predictions of HAB and hypoxia events may also mitigate impacts through advance warnings, providing time for treatment and avoidance strategies to be deployed.

There are numerous efforts in progress through the various Federal agencies that are designed to prevent, control, and mitigate HABs and hypoxia in the Great Lakes. For instance, the purpose of the GLRI is to target the biggest threats to the Great Lakes ecosystem, including strategies that lead to the reduction of HABs and hypoxia. A USGS study identifies nutrient sources and stream conditions that promote HABs and hypoxia by monitoring stream flow, nutrients, and sediment loads into the Great Lakes, providing information that will help to develop future prediction and mitigation strategies. Similarly, the USDA's Natural Resources Conservation Service (NRCS) partners with the USGS through the GLRI for monitoring water-quality improvements in streams along with edge-of-field monitoring. There has been a lot of progress in mitigation, as well as monitoring and modeling of nutrient input, which will lead to mitigation: tools to support prevention. The following sections include information on progress to-date towards addressing HABs and hypoxia in the Great Lakes.

The primary strategies for regulating HABs and hypoxia currently include the following:

- Minimizing land-based nutrients flowing into coastal and inland waters;
- Reducing nitrogen emissions in the airshed to reduce precipitation-related deposition of nitrogen compounds from the air to the land;
- Aerating systems and flushing dams to prevent hypoxia; and
- Reducing new introductions of algal species through genetic control (genetically engineering species that are purposely introduced into an ecosystem in order to alter the environmental tolerances, reproduction, or other processes of undesirable species) and environmental control (manipulation of an ecosystem environment, such as altering ecosystem habitat to disfavor growth of undesirable species.

8.5.1. Prevention

Prevention strategies in the Great Lakes regulate and examine the factors known to contribute to the development of HABs and hypoxia. This section reviews some Federal efforts to prevent HABs and hypoxia in the region.

The USDA's Agricultural Research Service (ARS), in partnership with NRCS, is also conducting edge-offield monitoring of conservation practices in numerous locations in the Western Lake Erie Basin under both USDA's CEAP Watershed Assessment Studies (including instream monitoring in the St. Joseph River Watershed in Indiana) and under the NRCS Conservation Innovation Grants Program. Water-quality monitoring is designed to detect changes as a result of conservation practices. CEAP Watersheds studies also examine how conservation practices interact and how collectively affect water quality in small watersheds. In addition, from 2008 through 2011 three other CEAP watershed studies in the Western Lake Erie Basin conducted by NRCS and USDA's National Institute of Food and Agriculture (NIFA) helped build understanding of conservation and water quality in the region.

USDA NRCS' CEAP Cropland Assessment conducted an extensive survey and modeling study in 2011 on soil conservation and nutrient reduction in the Great Lakes region. The assessment found that conservation practices between 2003 and 2006 significantly reduced in-stream loads of phosphorus, nitrogen, and sediment—by 20 percent, 21 percent, and 12 percent, respectively. The assessment also evaluated remaining conservation-treatment needs in the region. This information is being used by USDA, and state and other Federal agencies to improve conservation-program delivery to address HAB and hypoxia mitigation in the Great Lakes (USDA, 2011).

In 2012, USDA's NRCS and National Agricultural Statistics Service (NASS) administered a CEAP Cropland survey focused on the Western Lake Erie Basin, in order to examine additional conservation implementation in that region and changes in agricultural management. A detailed assessment of conservation effects is being conducted by USDA using that extensive survey data as well as other sources to compare with prior treatment and determine progress in the region as well as identify remaining conservation concerns. That assessment report is forthcoming and its findings will help to inform future conservation programming and planning to address the remaining needs within the Western Lake Erie Basin.

Modeling predicts that forage and cover crops, which take up nutrients from surface waters as opposed to subsurface waters, reduce both nitrogen and phosphorus losses in runoff (Francesconi et al., 2014). This is especially relevant in the Great Lakes region, which has some of the most productive agricultural soils in the world, but also would be largely unable to support agriculture without drainage. Drainage (e.g., tile drains) removes excess water that may otherwise prevent farmers from conduct field operations or would inhibit crop growth, but it can lead to the export of agrochemicals and nutrients through another pathway. Since 2008, USDA ARS scientists have monitored surface and tile discharges from agricultural fields in the St. Joseph River watershed. In contrast to the traditional understanding of tile drainage, which that suggests materials move slowly through soil to the tiles, USDA research shows that under certain specific conditions, peak water flow through the tiles sometimes occurs at the same time as peak surface runoff, suggesting a strong surface connection through the soil (D.R. Smith et al., 2014).

As part of the CEAP, scientists working in the St. Joseph River Watershed compared the effects of two conservation practices (no-till and reduced-till agriculture) on water quality. No-till reduced losses of all water-quality pollutants except soluble phosphorus (Smith et al., 2015). This can be the case when soils have high levels of phosphorous already present from historic management or are stratified at the surface, where there is tile drainage, or where soil macropores have developed over time. An APEX modeling study, conducted in collaboration with the Greater Wabash River Resource

Conservation and Development Council at the Wildcat Creek Watershed in Indiana, evaluated how the incorporation of additional conservation practices might be used to improve water quality. These studies highlight the importance of treating both surface runoff and tile drainage to minimize HABs.

In the young glacial till landscapes characteristic of the Great Lakes region, closed depressions known locally as potholes are pervasive. Water from surface drainage collects at the lowest spot in the pothole, keeping the area too wet for farming. Most farmed potholes are drained using subsurface tiles, but some also have supplemental drainage from a tile riser (a pipe with holes drilled in its sides) that extends vertically above the soil surface. ARS scientists in West Lafayette, Indiana, found that the extent of potholes within a watershed was directly related to concentrations or loads of nutrients lost from that watershed. USDA ARS research in a CEAP Watershed Assessment Study showed that an alternate conservation practice, called a blind inlet, provided greater filtration of surface water from potholes, decreasing soluble phosphorous losses by more than 50 percent, total phosphorus loss by 50 to 79 percent, sediment losses by up to 79 percent, and nitrogen loss by approximately 50 percent. In 2012, these ARS scientists worked with the NRCS to develop a modified conservation practice standard (modification to CP 620, Underground Outlet) that is now offered through the Environmental Quality Incentives Program (EQIP) in Indiana and Ohio and as an interim practice in some other states.

Extensive efforts to synthesize modeling data are underway through "Annex 4 – Nutrients Subcommittee" of the GLWQA. As part of this effort, Lake Erie has been elevated to the highest priority due to elevated nutrient levels and the large number of HABs being observed. In addition to intensive data-integration efforts, an ensemble modeling approach with nine models is being applied to different response indicators to meet the objectives of the Nutrients Annex.

The USGS SPARROW model helps to illustrate nutrient loading in rivers that flow into the Great Lakes. The resulting data can be used to further improve and test NOAA's HAB models. USGS has also created regional groundwater models and a major aquifer-sampling program. USGS projects include spatially relevant groundwater testing for nutrients that enter the Lakes through groundwater. Additionally, monitoring stations are located in small streams at the edge of agricultural fields where conservation practices are being implemented. The close proximity of the stream monitoring site to the field will allow for more rapid understanding of the effects of agricultural conservation practices on water quality. The study sites—which will run from 2014 through 2018—include the Fox River, which flows into Green Bay; a tributary to the Maumee River that flows into the western basin of Lake Erie; and a tributary to the Saginaw River that flows into Saginaw Bay. The results of the GLRI effort will be communicated to interested stakeholders to help guide similar efforts in the Great Lakes.

Data sharing has become an increasingly important component of research collaborations and is useful for connecting findings with interested stakeholders. The Great Lakes Observing System (GLOS) was created for this exact purpose. GLOS is a regional component of the Integrated Ocean Observing System that partners with Federal, regional, academic, and private entities to provide Great Lakes data to interested parties, including managers, decision-makers, and the public. NOAA and USGS are among the entities that provide data and products to GLOS.

NOAA's Great Lakes Environmental Research Laboratory (GLERL) has developed a HAB monitoring program for western Lake Erie that collects and distributes critical water-quality and toxicity data to regional water managers and other stakeholders on a weekly basis throughout the bloom season. These data are also used to validate the short- and long-term HAB forecasting models developed by NOAA's GLERL and National Centers for Coastal Ocean Service (NCCOS) scientists. The central Lake Erie buoy provides early warning to the Cleveland Water Department regarding dissolved oxygen levels and indications on movement of hypoxic water into drinking-water intakes. NOAA continues to expand and improve technologies and sampling efforts to determine conditions that promote HABs and hypoxia.

8.5.2. Control

EPA's Office of Water has developed analytical methods and advisories for HAB toxins in drinking water that are relevant to and applicable in the Great Lakes. This was in direct response to the 2014 Toledo drinking water closure, due to Lake Erie HABs. The agency is conducting water-treatment research to better understand the photochemical fate of toxins (i.e., the effect of light on toxins) and to test ultraviolet and solar-based methods of water treatment. The assessment includes efficacy of drinking-water treatment, cost-benefit analyses, and development of better testing methods. NOAA is developing natural filters for removal of toxins from drinking water and is doing a cost-benefit analysis of various HAB controls versus HAB impacts, for use in the Great Lakes and elsewhere.

8.5.3. Mitigation

Prediction and mitigation are interrelated. Prevention refers to acting before a HAB or hypoxic event occurs, while mitigation relates to the actions taken to respond to ongoing blooms. Mitigation efforts have the most immediate potential to reduce impacts, as they requires real-time management strategies and efforts that often are based on data accrued through prevention programs (Seltenrich, 2014).

The Great Lakes provide many examples of this relationship. Researchers at NOAA and the Cooperative Institute for Limnology and Ecosystems Research[4] in Ann Arbor, Michigan, have created the Lake Erie Operational Forecast System (LEOFS)[5] to predict HABs. The results of this model are published regularly by NOAA as a part of the Experimental Lake Erie Harmful Algal Bloom Bulletin.[6] The LEOFS predicted a bloom three days before the August 2014 Lake Erie bloom that affected the Monroe, Michigan, water-treatment plant. With information from the model, the treatment plan managers were better able to prepare for the influx of toxic water, thereby mitigating the human health effects of the HAB.

USDA, through the CEAP Watershed Assessment Studies, has conducted four small watershed-scale studies on the effects of conservation practices on water quality, water availability, and soil quality throughout the Great Lakes region. The goals of these projects are to document measureable water-quality (and related) changes as a result of the implementation of conservation practices, and to develop improved understanding of how conservation practices interact within a small watershed. USDA NRCS partnered with USDA ARS and USDA NIFA to conduct this work from 2004 to the present. One study[7] is ongoing in the St. Joseph River Watershed (Cedar Creek) in Indiana. Many important insights for more-effective conservation were identified from these studies and are now being used by NRCS in improved modeling algorithms, high-quality water-quality data, more-effective conservation practice standards, more-effective watershed and conservation planning, and comprehensive systems of practices. Each of these studies is being applied through conservation programs like the new RCPP and others in the region (particularly in the western Lake Erie Basin), where the studies were all conducted.

Agencies are working with Federal, state, and local officials, and other stakeholders, to develop better, faster responses to HABs. The CDC's National Outbreak Reporting System (NORS)[8] allows states to report human and animal sickness and deaths related to HABs and HAB toxins. To improve disaster preparedness, the CDC—with input from state, local, territorial, and tribal health departments—created a Community Assessment for Public Health Emergency Response (CASPER) tool. A CASPER survey was conducted by the Health Studies Branch of the National Center for Environmental Health (NCEH) at the CDC during the Toledo drinking water ban in 2014 to assess impacts of the Lake Erie bloom on Toledo residents (Toledo-Lucas County Health Department, 2014). The NCEH also helps with emergency-response preparedness and CASPER training for local, state, tribal, and territorial health departments. In 2014, NOAA NCCOS helped states respond to western Lake Erie HABs by doing toxin analysis in collaboration with labs at State University of New York – College of Environmental Science and Forestry, as well as by developing response plans and outreach materials.

9. CONCLUSION

Marine and freshwater HABs and hypoxia present serious challenges to protecting human, economic, animal, and environmental health across the United States. The direct economic effects of HABs alone in American communities include losses in income from commercial fishing, recreation, and tourism industries; public health costs of illness; expenses for monitoring and management; increased costs for drinking water treatment; and declining property values (Environmental Protection Agency, 2015). This amount does not include the costs of preventing or mitigating the causes of these events, which can run into the millions of dollars. As complex scientific processes, HABs and hypoxia present unique challenges to researchers for determining related causes, impacts, and best management practices. HABs and hypoxia also create social costs, including short- and long-term impacts on community development and social structure.

Although HABs and hypoxia occur naturally, it is clear that humans influence the severity and frequency of these events. The recommendations in this report are a roadmap for protecting American citizens and the country as a whole from this growing threat. The IWG-HABHRCA will use this action strategy to produce implementation plans, including timelines, for addressing individual components of the recommendations. Finally, the group will continue to engage with stakeholders around the country to determine how interagency efforts can be most responsive to evolving concerns and community needs related to HABs and hypoxia.

Appendix 1. HABs, Toxins, and their Effects

HAB Taxa	Toxin/Bioactive Compound	Human Health Effects	Animal Impacts	Environmental Effects	Economic Impact	Impacted Areas in U.S.
FRESHWATER						
Cyanobacteria	Microcystins, Cylindrospermopsin, Anatoxin-a, Saxitoxin(s), geosmins, methylisoborneol	Liver and kidney toxicity, neurotoxic, paralysis, gastrointestinal, dermatitis	Dog, farm animals, wildlife, and pet mortality, fish kills	Water discoloration, foul odors	Loss of tourism, contamination of drinking water requiring additional expensive water treatment or alternate water sources, make farmed and wild-caught freshwater fish inedible (including bad taste)	Great Lakes and many inland water bodies
Haptophytes (e.g., Prymnesium parvum, Chrysochromulina polylepis)	Prymnesins, Ichthyotoxins		Fish kills	Water discoloration, foam formation	Loss of fishing income, cleanup costs	TX,PA, OK, NM, AK, AL, CO, GA, NC, SC, WY, FL and NE
Chlorophytes Microalgae (e.g., Volvox, Pandorina)				Discolored water, localized hypoxia		Small eutrophic ponds

HAB Taxa	Toxin/Bioactive Compound	Human Health Effects	Animal Impacts	Environmental Effects	Economic Impact	Impacted Areas in U.S.
Macroalgae (e.g., Cladophora)		Bad odor	Associated with outbreaks of avian botulism	Fouls water intakes and piles up on beaches, localized hypoxia	Loss of recreational use and clean-up costs, clogged water intakes, bad odor	Great Lakes except Lake Superior, FL inland lakes
Euglenophytes (Euglena sanguinea)	Euglenophycin	Not characterized	Fish kills	Water discoloration	Loss of aquaculture operations	Texas, North Carolina, Great Lakes
Dinoflagellates (e.g., Peridinium polonicum [syn. Peridiniopsis polonicum, Glenodinium Gymnodinium])			Fish kills	Water discoloration		Discolored water in OK
Diatoms	Didymosphenia geminata			Produces large amounts of extracellular stalk material that piles up on beaches	Loss of recreational use	Nuisance blooms found in streams and rivers
MARINE						
Pseudo-nitzschia	Domoic Acid	Gastrointestinal and central nervous system effects (Amnesic Shellfish Poisoning)	Sea bird and marine mammal mortality		Closure of shellfish harvesting, loss of tourism income, wildlife rehabilitation	West Coast, Florida, Maine

Appendix 1. (Continued)

HAB Taxa	Toxin/Bioactive Compound	Human Health Effects	Animal Impacts	Environmental Effects	Economic Impact	Impacted Areas in U.S.
Dinophysis; Prorocentrum	Okadaic acid; Dinophysotoxins	Gastrointestinal distress			Closing of shellfish fisheries	Oregon, Texas, Washington
Gambierdiscus; Prorocentrum; Ostreopsis	Ciguatoxins	Sensory and gastrointestinal dysfunction (Ciguatera Fish Poisoning)	Possible marine mammal illness		Bans on fish sales from affected areas, cost of medical treatment, loss of protein source	Florida, Gulf Coast, Hawaii, Pacific, and Caribbean
Karenia	Brevetoxins	Neurotoxicity: Gastrointestinal and sensory effects (Neurotoxic Shellfish Poisoning), respiratory effects	Fish kills; manatee, dolphin, marine turtle, and Bird deaths	Water discoloration	Loss of tourism income; Removal of dead fish from beaches, shellfish fisheries closing, increased emergency room visits due to respiratory and gastrointestinal illness, wildlife rehabilitation	Gulf of Mexico and Atlantic coast up to North Carolina

HAB Taxa	Toxin/Bioactive Compound	Human Health Effects	Animal Impacts	Environmental Effects	Economic Impact	Impacted Areas in U.S.
Alexandrium; Gymnodinium; Pyrodinium bahamense	Saxitoxins	Respiratory paralysis, death (PSP)	Marine mammal deaths		Loss of shellfish harvesting income; human illness from recreational harvest; closure recreational puffer fish harvest in FL	Pacific coast, including Alaska; NE Atlantic coast; Florida
Karlodinium	Karlotoxins	Fish toxin	Fish kills	Water discoloration		Atlantic and Gulf Coasts
Aureoumbra lagunensis--Texas Brown Tide	Not characterized			Water discoloration; Loss of submerged aquatic vegetation		Texas, Florida
Aureococcus anophagefferens--Long Island Brown Tide	Not characterized			Water discoloration; seagrass and shellfish die- offs; hypoxic zones	Loss of shellfish harvesting income; interference with restoration	Mid-Atlantic coast
Akashiwo sanguineum	Surfactants	Suspected respiratory irritant	Migratory bird deaths, including protected species	Water Discoloration; Foam formation	Costs rehabilitation of protected bird species	Pacific Coast

Appendix 1. (Continued)

HAB Taxa	Toxin/Bioactive Compound	Human Health Effects	Animal Impacts	Environmental Effects	Economic Impact	Impacted Areas in U.S.
Heterosigma akashiwo	Ichthyotoxins		Fish kills	Water discoloration	Loss of fish net pen mariculture	Washington; Mid-Atlantic coast
Other Raphidophytes: Chattonella, Fibrocapsa	Brevetoxins; Ichthyotoxins		Fish kills	Water discoloration		Mid-Atlantic coast
Prorocentrum minimum-- Mahogany Tides	Not characterized		Mortality of spat in shellfish hatcheries	Water discoloration	Loss to shellfish hatcheries	Chesapeake Bay
Alexandrium monilatum	Goniodomin		Fish and shellfish mortality	Water discoloration		Gulf of Mexico and Atlantic coast up to New Jersey
Cochlodinium	Not characterized		Fish kills	Water discoloration	Severe impacts on fish and shellfish mariculture in Asia	West Coast, Mid-Atlantic
Macroalgae	H2S, dopamine	Respiratory impacts	Impair nesting protected species	Shade submerged Aquatic vegetation, cause hypoxia	Removal from beaches; loss of tourism income	All coasts
Marine cyanotoxins (formerly *Lyngbya*)	Dermatotoxins	Impact divers on coral reefs		Overgrow coral reefs	Loss of tourism income	South Florida

APPENDIX 2. HAB-RELATED HUMAN ILLNESSES

Toxin	Vector	Occurrence	Acute Toxicity	Long-Term Health Impacts	Susceptible Populations	Susceptible Regions
Anatoxin-a	Drinking water, Recreational waters, Dietary supplements	Low	Tingling, burning, numbness, drowsiness, incoherent speech, respiratory paralysis leading to death	Unknown	Recreational water users	Great Lakes, Continental U.S. ponds and lakes
Brevetoxins	Marine Aerosols, Recreational waters	High	Acute eye irritation, respiratory distress, asthma exacerbation	Unknown	Lifeguards, beachgoers, coastal residents	Gulf of Mexico
Brevetoxins	Shellfish	Low	Neurotoxic Shellfish Poisoning: Nausea, vomiting, diarrhea, numbness of lips, tongue, and throat, muscular aches, fever, chills, muscle pains, reduced heart rate, and pupil dilation	Unknown	Recreational shellfish harvesters	Gulf of Mexico, southeastern Atlantic coast

Appendix 2. (Continued)

Toxin	Vector	Occurrence	Acute Toxicity	Long-Term Health Impacts	Susceptible Populations	Susceptible Regions
Ciguatoxins	Fish	High	Ciguatera Fish Poisoning: Abdominal pain, nausea, vomiting, diarrhea; paresthesia, temperature dysesthesia, pain weakness; and bradycardia, hypotension	Recurrent symptoms of months to years, including neurological and neurophysiological symptoms (e.g., malaise, depression, headaches, myalgia, and fatigue)	Islanders subsistent on local fisheries, tourists, anglers in endemic areas	Florida Keys, Caribbean, Hawaii, Pacific Islands, Gulf of Mexico (Flower Garden Banks)
Cyanobacteria l LPS	Drinking water, Recreational Waters	Medium	Abdominal pain, vomiting and diarrhea, acute dermatitis	Unknown	Recreational water users	Great Lakes, Continental U.S. ponds and lakes
Cylindosperm-ospins	Drinking water, recreational waters, dietary supplements	Medium	Abdominal pain, vomiting and diarrhea, liver inflammation and hemorrhage, acute pneumonia, acute dermatitis	Malaise, anorexia, liver failure leading to death	Children, dialysis patients, liver disease, recreational water users	Great Lakes, Continental U.S. ponds and lakes

Toxin	Vector	Occurrence	Acute Toxicity	Long-Term Health Impacts	Susceptible Populations	Susceptible Regions
Domoic acid	Shellfish	Low	Amnesic Shellfish Poisoning: Vomiting, diarrhea, abdominal pain; confusion, disorientation, memory loss	Anterograde memory deficit, temporal lobe epilepsy, kidney disease	Elderly, children, pre-existing renal disease, subsistence and recreational harvesters	Northwest, California, Northeast, Gulf of Mexico
Microcystins	Drinking water, recreational waters, dietary supplements	High	Abdominal pain, vomiting and diarrhea, liver inflammation and hemorrhage, acute pneumonia, acute dermatitis, seizures leading to coma and death	Hepatocellular carcinoma, liver failure leading to death	Children, dialysis patients, pre-existing liver disease, recreational water users	Great Lakes, Continental U.S. ponds and lakes
Okadaic acid/ Dinophysistoxins	Shellfish	Low	Diarrhetic Shellfish Poisoning: Nausea, vomiting, diarrhea, abdominal pain accompanied by chills, headache, and fever	Gastrointestinal tumor promotor	Recreational shellfish harvesters	Northeast, Gulf of Mexico, Northwest

Appendix 2. (Continued)

Toxin	Vector	Occurrence	Acute Toxicity	Long-Term Health Impacts	Susceptible Populations	Susceptible Regions
Saxitoxins	Shellfish	Medium	Paralytic Shellfish Poisoning: tingling, burning, numbness, drowsiness, incoherent speech, respiratory paralysis leading to death	Unknown	Children, recreational harvesters	Alaska, Northwest, Northeast, California, Northeast Florida
Yessotoxins, Pectenotoxins	Shellfish	Unknown	Not documented as toxin in humans, but co-occur in DSP shellfish and are highly toxic to mice	Unknown	Unknown	Northeast, Northwest

Appendix 3. Agency Activities on HABs and Hypoxia

Office/ Dept.	Agency	HABs/ Hypoxia/Both	Program Title (brief description)	Program Activities
DHHS	CDC	HABs	National Outbreak Reporting System (NORS)	CDC initiated waterborne and foodborne disease outbreak surveillance systems in the 1970s. U.S. states and territories voluntarily report to these systems via the electronic NORS, which receives aggregate data on human cases and their exposures, including exposures to harmful algal blooms (HABs) or HAB toxins. The One Health Harmful Algal Bloom System (OHHABS) is being developed for single case-level reporting of human and animal illness, and relevant environmental data. OHHABS is being programmed to inform restoration activities in the Great Lakes but will accessible to all states via NORS. The pilot version of the system is being tested in preparation for a 2016 launch.
DHHS	CDC	HABs	Great Lakes State Health Surveillance Capacity	CDC has partnered with the Council of State and Territorial Epidemiologists (CSTE) since 2013 to place and provide technical support for epidemiology fellows in Great Lakes states, including Indiana, Illinois, Michigan, Minnesota, New York, Ohio, and Wisconsin. The activity is supported by the Great Lakes Restoration Initiative. Fellows focus on waterborne disease detection, investigation, response and reporting.

Appendix 3. (Continued)

Office/ Dept.	Agency	HABs/ Hypoxia/Both	Program Title (brief description)	Program Activities
				The fellowship has expanded state waterborne disease reporting and analytic capacity; improved state health surveillance for harmful algal blooms; and ensured dedicated staff time for waterborne disease surveillance and coordination activities.
DHHS	CDC	HABs	Health Communications	CDC's health communications activities related to HABs include the preparation of a HAB website with information for public health practitioners, clinicians, and the general public, and the expansion of the Drinking Water Advisory Communications Toolbox (DWACT) to include information about HAB-related drinking water advisories. The DWACT was created through a collaborative effort among CDC, EPA, the American Water Works Association, the Association of State and Territorial Health Officials, the Association of State Drinking Water Administrators (ASDWA), and the National Environmental Health Association (NEHA).
Multiple	CDC, EPA, NOAA	HABs	Interagency Analytic Workgroup	Additional research is needed to fully characterize and understand the health risks from drinking water provided by public water systems when that water is contaminated with cyanobacterial toxins.

Office/ Dept.	Agency	HABs/ Hypoxia/Both	Program Title (brief description)	Program Activities
				There is a need to establish standardized biological sample collection and analysis protocols to support assessment of toxin-associated health effects. Multiple federal agencies are working together to assess sampling and analytical capabilities related to analysis of biological specimens collected from human and animals exposed to cyanobacteria toxins via contaminated water, including drinking water. The goal is to combine expertise to develop robust analytic methods to detect biological evidence of exposure to cyanobacterial toxins, to optimize laboratory and emergency response capacity in the collection, analysis, and response to harmful algal bloom-related illnesses.
DHHS	CDC	HABs		Method development, refinement, and validation for detecting human exposures to HAB toxins through the detection of toxins and specific biomarkers in clinical samples. Current methods approved for use include the detection of saxitoxin, neosaxitoxin, tetrodotoxin, and gonyautoxins (1-4), which have been applied to individual cases to confirm suspected HAB exposures.
DHHS	FDA	HABs		Method development, refinement, and validation for detecting HAB toxins; Improving understanding of HAB toxin sources and vectors that impact seafood and dietary supplement safety

Appendix 3. (Continued)

Office/ Dept.	Agency	HABs/ Hypoxia/Both	Program Title (brief description)	Program Activities
DHHS	FDA	HABs		Developed, evaluated, and validated rapid screening for HAB toxins in seafood, thereby improving regulatory monitoring, surveillance programs, and outbreak response. For example, FDA developed an onboard screening dockside testing program for PSP toxins in shellfish, which led to reopening of a large portion of Georges Bank in 2013 to safe commercial harvest of clams.
DOC	NOAA	HABs		The NOAA Northwest Fisheries Science Center has established coastal monitoring program called ORHAB (funded by MERHAB) for outer WA coast monitoring and a new monitoring partnership called SoundToxins for the early warning of harmful algal blooms in Puget Sound. The NOAA National Centers for Coastal Ocean Science Ecology and Oceanography of Harmful Algal Blooms (ECOHAB) program provided 3 years of funding to develop the Puget Sound Harmful Algal Bloom (PS-AHAB) project to understand environmental controls on the benthic (cyst) and planktonic life stages of the toxic dinoflagellate *Alexandrium*, and evaluate the effects of climate change on the timing and location of blooms.

Office/ Dept.	Agency	HABs/ Hypoxia/Both	Program Title (brief description)	Program Activities
				ECOHAB funded a 3-year project on the ecology of *Heterosigma akashiwo*, a fish-killing flagellate, that has resulted in millions of dollars lost to net-penned salmon in Puget Sound.
DOC	NOAA	HABs	National Phytoplankton Monitoring Network	The PMN was established to monitor phytoplankton and harmful algal blooms and promote environmental stewardship though the use citizen volunteers. PMN volunteers are trained by NOAA staff on sampling techniques and identification methods for over 50 genera, including 10 potentially toxin producing genera, of dinoflagellates and diatoms on the volunteers watch list. Currently, 250 sites in 22 states and American territories including 52 schools, 15 universities, 298 civic groups and 40 state and Federal agencies collect phytoplankton and environmental data. Since the inception of the program in 2001, more than 275 algal blooms and 15 toxic events have been reported by PMN volunteers.
DOC	NOAA	HABs	National Analytical Response Team (ART)	NOAA's ART provides rapid and accurate identification and quantification of marine algal toxins in suspected HABs, and related marine animal mortality events and human poisonings.

Appendix 3. (Continued)

Office/ Dept.	Agency	HABs/ Hypoxia/Both	Program Title (brief description)	Program Activities
				From 2009 to 2014, ART received over 4000 samples from state and Federal government agencies, non-governmental organizations, and academic partners for determination of toxins associated with harmful algae. In addition to water samples, marine toxins were analyzed in samples from marine and freshwater algae, shellfish, fish, cetaceans, pinnipeds, birds and sea turtles. ART has analyzed samples from almost every US coastal state and the U.S. Virgin Islands and from countries including Canada, United Kingdom, Morocco, Argentina, Chile, Mexico, Uruguay, Peru, and El Salvador.
DOC	NOAA	HABs	Technology Transfer Team	The Technology Transfer Team completed rigorous, international inter-laboratory trials in partnership with interagency organizations, Federal agencies and private businesses to bring the receptor binding assay for PSP toxins to national and international regulatory approval; and guided its commercialization to assure American shellfish are safe for American citizens and export throughout the world.

Office/ Dept.	Agency	HABs/ Hypoxia/Both	Program Title (brief description)	Program Activities
				The team also provided training on use of the method to more than 30 countries through formal agreement with the International Atomic Energy Agency to promote safe food supply and increased economic growth through the export of fisheries products and to the Southeast Alaska Tribal partnership to enable monitoring of subsistence resources.
DOC	NOAA	HABs		The National Centers for Coastal Ocean Science (NCCOS) has provided base funding (labor and operational funds) for development of HAB forecasts; NOAA anticipates this to continue through fiscal year 2015. NOAA is starting new research projects on using satellite data to improve the frequency of the forecasts for the operational Gulf of Mexico (with funding by the NASA ROSES Health and Air Quality program) and in coordination with EPA, NASA, and USGS on freshwater systems. (This includes funding by NASA Ocean Biology program and tentative support through fiscal year 2016funding from the Great Lakes Restoration Initiative). NOAA is also working with EPA on systematic approach to either warning state health/water quality on cyanobacteria blooms and allowing them to evaluate patterns and trends in lakes and estuaries that are at risk.

Appendix 3. (Continued)

Office/ Dept.	Agency	HABs/ Hypoxia/Both	Program Title (brief description)	Program Activities
DOC	NOAA	HABs		Make satellite coverage of ocean and coastal zones more comprehensive and combine it with existing data to enable quantifiable estimates of HABs (much of this has been funded by NASA). Plan to transfer promising new monitoring and prediction technology and approaches from research to operational HAB forecasts for Gulf of Mexico, Lake Erie, Chesapeake Bay, Puget Sounds, Pacific Northwest, and California.
DOC	NOAA	HABs	Ecology and Oceanography of Harmful Algal Blooms (ECOHAB)	Developing a better understanding of HAB causes and impacts that form the basis for better management to reduce HABs and their impacts.
DOC	NOAA	HABs	Monitoring and Event Response for Harmful Algal Blooms (MERHAB)	National, competitive extramural research program that builds capacity for enhanced HAB monitoring and response in state, local, and tribal governments.
DOC	NOAA	HABs	Prevention, Control, and Mitigation of Harmful Algal Blooms (PCMHAB)	National, competitive extramural research program that develops new methods of HAB prevention, control, and mitigation. It also addresses the socioeconomic impact of HABs and efforts to reduce HAB impacts.
DOC	NOAA	HABs	Event Response	Provides immediate assistance for managing HAB events and advancing the understanding of HABs when they occur.

Office/ Dept.	Agency	HABs/ Hypoxia/Both	Program Title (brief description)	Program Activities
DOC	NOAA	HABs		NESDIS/STAR has provided support in FY15 to transfer VIIRS remote sensing data for GOMx HAB forecasts into operations. STAR is currently planning support for FY16+ to acquire and process remote sensing data (e.g., from Sentinel-3/OLCI) that meets HAB requirements for expanded geographic regions.
DOC	NOAA	Hypoxia	Coastal Hypoxia Research Program (CHRP)	National, competitive extramural research program that develops understanding of hypoxia causes and impacts that form the basis for better management to reduce hypoxia and its ecological and socioeconomic impacts. Includes all coastal systems except the large hypoxic zone along the northern Gulf of Mexico continental shelf.
DOC	NOAA	Hypoxia	Northern Gulf of Mexico Ecosystems and Hypoxia Assessment Program (NGOMEX)	National, competitive extramural research program that develops understanding of the causes and impacts of the northern Gulf of Mexico hypoxic zone, that form the basis for better management to reduce the hypoxic zone and its ecological and socioeconomic impacts.
DOC	NOAA	Hypoxia		Continue to convene workshops to obtain stakeholder needs that drive research prioritization, and disseminate advanced knowledge and tools for hypoxia mitigation to regional managers and interagency management networks such as the Gulf Hypoxia Task Force or the Landscape Conservation Cooperative.

Appendix 3. (Continued)

Office/ Dept.	Agency	HABs/ Hypoxia/Both	Program Title (brief description)	Program Activities
DOC	NOAA	HABs		Studies molecular ecology of HABs in the Great Lakes and makes improvements to monitoring HABs and toxicity in the Great lakes. Monitors six routine stations in the western basin of Lake Erie weekly during blooms season and supplies data that supports the predictive models in Lake Erie. Developing a three-dimensional lagrangian particle transport model to effectively predict HAB advection as part of the Lake Erie Operational Forecasting System, which is set to go operational in fiscal year 2015.
DOD	USACE	HABs		Responding to HABs in response to public reports/complaints in close coordination with State water quality/public health agencies. Response programs developed by individual USACE Divisions/Districts. USACE Engineer Research and Development Center available to support Divisions/Districts in assessing HAB impacts to USACE Civil Works Projects (e.g., water quality modeling, remote sensing, and technical assistance). General water quality monitoring and HAB response to meet authorized project purposes and recreation mission requirements.

Office/ Dept.	Agency	HABs/ Hypoxia/Both	Program Title (brief description)	Program Activities
DOD	Navy	HABs		Studies the variability of *in situ* and remotely sensed spectral optical properties to identify dinoflagellates through field sampling and improvement of remote sensing techniques. Dinoflagellate information has been incorporated into Naval Research Laboratory's ecological-circulation models for better understanding/prediction.
DOI	Bureau of Ocean Energy Management (BOEM)	Hypoxia	Gulf Oxygen Deepwater Experiment	Study plan pending approval: To address noted data gaps that addresses deepwater oxygen dynamics such as in the Oxygen Minimum Zone in the Gulf, whereas the Louisiana-Texas Shelf Physical Oceanography Program shelf studies were on the shallower hypoxic zone.
DOI	NPS	HABs and hypoxia	Outreach and Education	Of the 407 NPS units, there are 86 units that are considered ocean, coastal, or Great Lake parks, in addition to other park units that have extensive surface water bodies. HABs have the potential to influence all of these park units at various levels, and it is therefore important to prepare for these events in order to preserve our resources. The National Park Service is creating a website containing a public health and ecological HAB events reporting system. It also provides a point of contact for park managers to partner with local and state health and environmental agencies who will provide park personnel with technical assistance for the management of HAB events.

Appendix 3. (Continued)

Office/ Dept.	Agency	HABs/ Hypoxia/Both	Program Title (brief description)	Program Activities
				Outreach materials (brochures, interpretive displays, and materials) on HABs, their causes, the effects on the ecosystem, and the many ways to reduce or stop nonpoint source pollution, many of which are simple to implement.
DOI	USGS	HABs and hypoxia	National Water Quality Program	USGS conducts long-term monitoring of nutrients and other water quality characteristics in surface and groundwater networks. The sources and quantities of nutrients delivered by streams and groundwater to the Great Lakes and estuaries are monitored at 106 sites. Annual updates from the monitoring sites will be made available to the public, including nutrient concentrations, loads, and yields. These data, along with data aggregated from numerous other agencies, are used to evaluate trends in critical water quality parameters including nutrients and sediment. Real-time measurements for dissolved oxygen and temperature are collected at over 500 and 2000 locations, respectively. USGS is pioneering new field sensor methods and systems for monitoring and delivering real-time nutrient data. The USGS SPARROW model quantifies nutrient sources and sediment loads to coastal areas, the Great Lakes, and inland lakes in the Eastern United States.

Office/ Dept.	Agency	HABs/ Hypoxia/Both	Program Title (brief description)	Program Activities
				SPARROW has also been linked to an online Decision Support System, which allows direct exploration of the potential benefits of nutrient management for systems including the Chesapeake and the Mississippi, other coastal rivers, and the Great Lakes.
DOI	USGS	HABs and hypoxia	National Water Quality Program/National Water Quality Assessment	USGS collects fish-, aquatic macroinvertebrate-, and algae-community samples, and conducts stream physical habitat surveys to assess the effects of multiple stressors—including algal toxins—on aquatic organisms in streams in several ecoregions.
DOI	USGS	HABs	National Water Quality Program/ Cooperative Water Program	HAB research is conducted in at least 20 USGS Water Science Centers. Studies include both short- and long-term projects focused on quantifying blooms and associated toxins and taste-and-odor compounds, and understanding causal factors. Many studies employ new and developing sensor technology to detect algal pigments. For example, a study of the primary drinking water supply for Wichita, Kansas combined long-term discrete and continuous water-quality data to develop models that estimate the probability of microcystin occurrence in near real time.
DOI, USDA	USGS, USDA-NRCS	HABs and hypoxia	GLRI	GLRI assesses the impacts of agricultural management practices, climate change, and land use change on the timing and magnitude of delivery of nutrients and sediments to the Great Lakes at 30 sites.

Appendix 3. (Continued)

Office/ Dept.	Agency	HABs/ Hypoxia/Both	Program Title (brief description)	Program Activities
				The group works with NOAA, EPA, states, universities, and NGOs to understand how nutrient and sediment loading from the Great Lakes watershed affect hypoxia, HABs and biological communities in the near-shore environment. Edge-of-field studies in the GLRI priority watershed quantify phosphorus, nitrogen, and sediment to evaluate nutrient reduction projects on agricultural land. Rapid sharing of edge-of-field monitoring results with local stakeholders allow for adaptive implementation.
DOI	USGS	HABs	Toxic Substances Hydrology Program	The Toxic Substances Hydrology Program supported research investigates the origins, occurrence, transport/fate, effects, and mitigation of HABs and associated toxin mixtures. In doing so new methods are being pioneered including toxin-specific analytical methods and development of targeted and non-targeted ground-to-space field and laboratory methods. Current and planned research, which includes investigations of metabolites and related biota, characterizes the spatial/temporal extent of understudied aspects of toxins associated with HABs; evaluates environmental controls

Office/ Dept.	Agency	HABs/ Hypoxia/Both	Program Title (brief description)	Program Activities
				responsible for HAB proliferation and associated toxin production; and evaluates environmental health implications and impacts. In addition to contributing to basic understandings of the biogeochemical underpinnings of algal toxin occurrence and associated environmental health threats, this information can be utilized to assist with standardization of study designs, field, laboratory, and interpretative techniques and to inform mitigation activities. Current and planned collaboration is ongoing with multiple federal and state agencies and tribes as well as through outreach efforts such as participation on the Inland HAB Discussion Group. Industry collaborations are fostered to facilitate acquisition of lower cost, higher throughput screening assays and more advanced interpretative capabilities where TSHP provides validation support for the benefit of program research and stakeholder collaboration.
DOI	USGS	HABs	Energy, Mineral, and Environmental Health/Toxic Substances Hydrology Program	Pioneer new field monitoring methods (sensors), assessment techniques, and laboratory methods needed to address harmful algal bloom issues in freshwaters. New methods include a multi-toxin method that can quantify cyanotoxin mixtures, and DNA- and RNA-based molecular methods for detecting microcystin and microcystin producers.

Appendix 3. (Continued)

Office/ Dept.	Agency	HABs/ Hypoxia/Both	Program Title (brief description)	Program Activities
DOI	USGS	HABs	Ecosystems	USGS has ongoing research characterizing ecological and food web impacts of cyanotoxins. For example, a USGS study in Upper Klamath Lake demonstrated a link between microcystin and reduced young-of the year recruitment of federally endangered suckers.
NSF	Joint initiative between NSF and NIEHS	HABs	Oceans and Human Health (OHH) Initiative and the NSF's Division of Ocean Sciences	The NIEHS supports multiple studies focused on the effects of HAB toxins on human and mammalian physiology, development of biomarkers for chronic toxin exposure, and the design and testing of novel technologies for *in situ* detection of algal toxins in fresh and salt water environments. For example, a number of ongoing studies are supported that analyze the effects of domoic acid on neurotoxicity as well as cognitive impacts in human cohorts, non-human primates and rodent models. Also, NIEHS is accepting unsolicited applications for support and use of time sensitive mechanism to allow research support for unanticipated bloom events.
NSF	NSF	HABs and hypoxia	Ocean Observatories Initiative and the National Ecological Observatory Network	Provides environmental data for studies of HABS (both marine and freshwater) and hypoxia.

Office/ Dept.	Agency	HABs/ Hypoxia/Both	Program Title (brief description)	Program Activities
NSF	NSF	HABs	Division of Ocean Sciences (OCE), NSF Ocean Observatories Initiative	Observational capabilities for research in marine systems.
NSF	NSF	HABs	Directorate for Geosciences, Prediction and Resilience Against Extreme Events (PREEVENTS)	Focused interdisciplinary research projects.
NSF	NSF	HABs	Division of Biological Infrastructure, National Ecological Observatory Network (NEON)	Observational capabilities for ecological research.
NSF	NSF	HABs	Division of Ocean Sciences	Research Support, unsolicited proposal in marine ecology.
NSF	NSF	HABs and hypoxia	Collaboration between NSF GEO, SBE, and ENG directorates, as well as USDA NIFA.	Program supporting interdisciplinary research to understand and predict the interactions between the water system and climate change, land use, the built environment, and ecosystem function and services though research and models. Several research projects are focused on nutrient movement and hypoxia mitigation strategies.

Appendix 3. (Continued)

Office/ Dept.	Agency	HABs/ Hypoxia/Both	Program Title (brief description)	Program Activities
NSF	NSF	HABs	Ocean and Human Health Initiative, a Collaboration between NSF's Division of Ocean Sciences, and the National Institute for Environmental Health Sciences	Studies of the effects of HAB toxins on human and mammalian physiology, development of biomarkers for chronic toxin exposure, and the design and testing of novel technologies for *in situ* detection of algal toxins in fresh- and salt-water environments. Also accepting unsolicited applications for support and use of time sensitive mechanism to allow research support for unanticipated bloom events.
USDA	NIFA and ARS	HABs		Support of extramural and intramural research on the effects of HABs and HAB toxins on food safety, aquaculture, and livestock.
USDA	ARS	HABs and hypoxia		Research on nutrient management, nutrient contribution to hypoxia, and aquaculture. Long-Term Agro-Ecosystem Research (LTAR) and Watershed Research Centers.
USDA	NIFA	HABs and hypoxia		Research support for studies of the effects of nutrient cycling, climate change, and nutrient management for agriculture.

Office/ Dept.	Agency	HABs/ Hypoxia/Both	Program Title (brief description)	Program Activities
USDA	NRCS	HABs and hypoxia	Conservation Technical Assistance (planning); Environmental Quality Incentives Program; Conservation Stewardship Program; Agricultural Conservation Easement Program Regional Conservation Partnership Program	NRCS provides conservation planning assistance to agricultural producers on cropland, grazing land, and for confined livestock operations. NRCS also has financial assistance programs to help producers implement and install practices. These programs are all voluntary and are incentive-based. For confined livestock systems, this includes, but is not limited to, practices such as waste storage structures, and associated practices like roofs and covers, roof runoff management, diversions, and a nutrient management plan for the utilization of manure. On cropland, this may include agronomic practices such as residue management, cover crops, conservation cropping systems, and nutrient management; buffer practices like filter strips and riparian forest buffers; water management practices such as grassed waterways, grade stabilization structures, drainage water management, blind inlets (to replace surface inlets), wetland restoration and creation; and prescribed grazing systems and associated practices for grazing land. NRCS also assists farmers financially with edge-of-field water quality monitoring.

Appendix 3. (Continued)

Office/ Dept.	Agency	HABs/ Hypoxia/Both	Program Title (brief description)	Program Activities
USDA	NRCS	HABs and hypoxia	Great Lakes Restoration Initiative, Mississippi River Basin Healthy Watershed Initiative, National Water Quality Initiative, etc.	Under various water quality initiatives, NRCS and its partners help producers in selected watersheds to voluntarily implement conservation practices that avoid, control, and trap nutrient runoff; improve wildlife habitat; and maintain agricultural productivity. These initiatives utilize NRCS programs such as the Environmental Quality Incentives Program (EQIP) and the Conservation Stewardship Program (CSP) within targeted watersheds to provide technical and financial assistance.
USDA	NIFA and ARS	HABs and hypoxia		Supports research on best management practices for nutrient management, aquaculture, and plant breeding, among others. Specific concerns addressed by this research include manure management from animal feeding operations and water use and conservation on irrigated cropland.
USDA	NRCS, ARS (partnership with The Nature Conservancy)	HABs and hypoxia	CEAP— Wildlife — Western Lake Erie Basin	The on-going Nature Conservancy-led Western Lake Erie Basin CEAP-Wildlife project is being conducted to assess and forecast benefits of NRCS conservation practices to stream fish communities, to help advance strategic conservation of riverine ecosystems.

Office/ Dept.	Agency	HABs/ Hypoxia/Both	Program Title (brief description)	Program Activities
				A similar project for the Saginaw Bay was already completed. In this WLEB CEAP Wildlife project, using pre-existing water quality and stream fish community data, the effort is linking SWAT modeling with fish community condition at small watershed scales (NHD+ scale) to reveal relationships between conservation practice implementation and fish community response within the streams. The effort will provide science-based estimates of the priorities, scope and costs of restoring stream fish communities throughout the Western Lake Erie Basin watershed. Coordination with other groups is evaluating connections to the Lake Erie water quality.
EPA	EPA	HABs and hypoxia	Water Quality Management	Diversified approach to better understand cyanobacterial HABs ecology and the development of watershed and source water management techniques, including the development of models for nutrients loadings, the optimization of watershed placement of phosphorus and sediment BMPs, and the use of water quality trading (WQT) to cost-effectively reduce nutrient loadings. It also includes an assessment of the impact of land use and infrastructure on watershed changes, and the evaluation of ecological contributors to cyanobacterial HAB development and toxin production.

Appendix 3. (Continued)

Office/ Dept.	Agency	HABs/ Hypoxia/Both	Program Title (brief description)	Program Activities
				This research program also includes the use of molecular methods to characterize the risk in a reservoir for toxin and algal blooms, and the analysis of the impact of HABs on creating disinfection byproducts (DBPs) precursors.
EPA	EPA	HABs	Human and Ecological Health	Research support to address data gaps associated with health, ecosystem, and economic effects of HABs. Research activities include the characterization of cyanobacteria and their toxins and allergic components, the evaluation of the toxicity of multiple congeners of microcystins, and identification of biomarkers of exposure for human health risk assessments. Epidemiology studies to characterize toxin occurrence in U.S. inland lakes, and studies to determine that bioaccumulation, bioconcentration, and biomagnification of cyanotoxins in mammalian tissues and food web are also in place. EPA is also assessing occurrence and health information for the inclusion of cyanotoxins in the Contaminant Candidate List (CCL) and the Unregulated Contaminant Monitoring Rule (UCMR) program. In addition, EPA is developing Human Health Water Quality Criteria (HHWQC) for cyanotoxins in recreational waters.

Office/ Dept.	Agency	HABs/ Hypoxia/Both	Program Title (brief description)	Program Activities
EPA	EPA	HABs	Monitoring and Analytical Methods Development	A collaborative effort of EPA, NASA, NOAA, and USGS to provide an approach for mainstreaming satellite ocean color capabilities into U.S. fresh and brackish water quality management decisions. The Cyanobacteria Assessment Network (CyAN) for freshwater systems will develop approaches to relate nutrient loads and land use to the frequency, location, and severity of cyanobacterial blooms in lakes of the U.S. It will include assessing risk to human health from satellite multispectral data to assess biological conditions and risk to human health in lakes and reservoirs in the United States. EPA also provides nationally consistent and scientifically defensible assessments of aquatic resources through the National Aquatic Resource Surveys (NARS), including indicators associated with cyanotoxin exposure. EPA and its regions are also working on monitoring efforts such as the Lake Champlain Cyanobacteria Monitoring, Great Lakes Restoration Initiative projects and Phosphorus Reduction Strategy, Southeast Alaska Tribal Toxins (SEATT) project, and the Puget Sound Toxins Project.

Appendix 3. (Continued)

Office/ Dept.	Agency	HABs/ Hypoxia/Both	Program Title (brief description)	Program Activities
				EPA is also working on monitoring projects to improve identification and removal of HAB toxins in drinking water, and evaluating the impact of temperature on bloom development. EPA is developing analytical tools including the use of real-time sensors, quantitative polymerase chain reaction and fluorescence based technologies of micro spectrophotometer and flow cytometry to detect cyanobacteria organisms in source water.
EPA	EPA	HABs	Drinking Water Treatment	EPA is working collaboratively with regional offices to characterize the effectiveness of drinking water treatment techniques in reducing toxins.
EPA	EPA	HABs	Outreach	EPA conducts webinars and provides online resources to promote public awareness and information sharing.
NASA	NASA	HABs	The Ocean Biology and Biogeochemistry Program	Basic HABs research resulting in publications and new retrieval algorithms.
NASA	NASA	HABs	Health and Air Quality Applications Program	Improve the forecast resolution and frequency of risk of *Karenia brevis* toxins on every beach, every day, rather than every county, twice a week. The methods would be applicable across the Gulf of Mexico.

Office/ Dept.	Agency	HABs/ Hypoxia/Both	Program Title (brief description)	Program Activities
NASA	NASA	HABs	Health and Air Quality Applications Program	Monitoring and surveillance of cyanobacterial harmful algal blooms in drinking and recreational water supplies. Satellite derived products that were developed for western Lake Erie are being analyzed for their use in other regions (e.g., Chesapeake Bay and inland lakes in Ohio and Florida). This project has established methods to identify environmental thresholds that indicate the potential for cyanobacterial blooms to form or persist, and these data sets are also being made available to CDC.
	Multiple agencies and partners, including but not limited to EPA, FWS, NOAA, NPS, USACE, USDA, USGS	HABs and hypoxia	Water Quality Portal	Co-sponsors of the Water Quality Portal, a cooperative data service that makes data publically available. The data are derived from the USGS National Water Quality Information System (NWIS), the EPA Storage and Retrieval data warehouse (STORET), and the USDA ARS Sustaining the Earth's Watersheds - Agricultural Research Database System (STEWARDS). With data from over 400 Federal, state, tribal, and local agencies, this efforts will improve understand progress in nutrient reduction efforts.
	Multiple agencies: CDC, NASA, NOAA, NSF, USDA, and USGS	HABs	Exposure Science (ES)21 Federal Working Group on Exposure Science	Exposure assessment is instrumental in helping to forecast, prevent, and mitigate exposure that leads to adverse human health or ecological outcomes.

Appendix 3. (Continued)

Office/ Dept.	Agency	HABs/ Hypoxia/Both	Program Title (brief description)	Program Activities
				This vision expands exposures from source to dose, over time and space, to multiple stressors, and from the molecular to ecosystem level. HAB exposure assessment is addressed by ES21 Working Groups on Biomonitoring, Citizen Engagement/Citizen Science and Sensors/Dosimeters.
	Multiple Agencies, EPA and NOAA	HABS	Volunteer Freshwater Phytoplankton Monitoring Program	Volunteer monitoring program that collects baseline data on harmful algal species and builds capacity by providing data to NOAA Phytoplankton Monitoring Network and EPA. Volunteers are trained to identify algae, collect water samples, conduct basic water quality analyses, and preserve samples for further analysis NOAA Analytical Response Team. Network became operational in 2015 with stations in the Western Basin of Lake Erie in seven lakes in EPA Region 8 with plans to expand to Lakes Michigan, Superior, Huron and Grand Lake St. Mary in 2016.

Office/ Dept.	Agency	HABs/ Hypoxia/Both	Program Title (brief description)	Program Activities
USDA	USDA/Multiple agencies, led by USDA NRCS, ARS, NIFA, FSA, and NASS. Also includes USGS, NOAA, FWS, EPA, BLM, NASA, USDA Economic Research Service and US Forest Service	HABs and hypoxia	CEAP	The Conservation Effects Assessment Project (CEAP) is a collaborative, multi-agency effort to quantify the environmental effects of conservation practices and programs and develop the science base for managing the agricultural landscape for environmental quality. Project findings are used to guide USDA conservation policy and program development, and help conservationists, farmers, and ranchers make more informed conservation decisions. USGS will incorporate conservation data collected by CEAP into their surface water quality monitoring. The National Cropland Assessment combines information from NASS producer surveys and conservation practice data as inputs into two models [Agricultural Policy EXtender field-scale model and Hydrologic Unit Model for the United States/Soil and Water Assessment Tool (HUMUS/SWAT) watershed model] to estimate the environmental benefits of conservation practices and conservation treatment needs within major drainage basins of America. These include sub-basins of the Mississippi River Basin, Chesapeake Bay and Great Lakes. In addition, the Watershed Assessment Component of CEAP conducts small

Appendix 3. (Continued)

Office/ Dept.	Agency	HABs/ Hypoxia/Both	Program Title (brief description)	Program Activities
				watershed scale studies across the United States to quantify water and soil resource outcomes of conservation practices and systems and enhance understanding of processes. Interactions among practices are investigated as well as modeling enhancements, watershed targeting approaches, and socioeconomic factors. Practice standards are developed or updated to improve effectiveness and address gaps.
USDA	Multiple agencies, led by USDA NRCS, ARS, NASS and FSA	HABs and hypoxia	CEAP	In 2012, NASS worked with NRCS to administer a CEAP Cropland-survey focused on the Western Lake Erie Basin. Data from the survey and other sources is being used to assess conservation effects in the Western Lake Erie Basin and compare trends and progress in conservation as well as evaluate additional treatment needs in that region. The assessment report is forthcoming.

Appendix 4. Actions Taken Since 2008 HABHRCA Reports – HABs

Recommendation Categories	2008 HABHRCA Recommendations	Actions Taken	Agencies
Research	Effective prevention, control, and mitigation requires an understanding of the factors contributing to HAB events and knowledge of their impacts on ecosystems, wildlife, livestock, and humans	Determine causes and impacts of HABs	EPA, FDA, FWS, NIEHS, National Institute of Standards and Technology (NIST), NSF, NOAA, USDA, USGS
Prevention	Modification of nutrient inputs	Ongoing	EPA, USACE, USDA
	Modification of hydrodynamic conditions	Ongoing	USACE
	Minimize or prevent introductions of invasive HAB species, their cysts, and organisms that facilitate the success of HAB species	Ongoing	EPA, NOAA
Control	Eliminate or reduce the levels of HAB organisms and their toxins	Evaluate various options for destroying and/or removing HAB cells and toxins with physical, chemical, or biological methods hy environmentally and socially acceptable means	EPA, NOAA, USDA

Appendix 4. (Continued)

Recommendation Categories	2008 HABHRCA Recommendations	Actions Taken	Agencies
Mitigation	Improve monitoring of HAB cells and toxins	Develop sensitive, quantitative, field deployable assays and sensors for HAB cells, toxins, and relevant toxin metabolites. Develop remote sensing capabilities for HABs that can be utilized in monitoring and management programs. Integrate HAB and toxin detection technologies into emerging U.S. and global ocean observation systems.	EPA, FDA, NASA, NIEHS, NIST, NOAA, NSF, USACE, USGS
	Facilitate disease surveillance, clinical characterization, and therapeutic guidance in humans and animals	Develop and validate biomonitoring methods of toxin exposure and identify and utilize biomarkers for toxin exposure effect and disease status. Enhance disease surveillance for human and animal illnesses and deaths resulting from HAB toxin exposure by supporting existing surveillance systems.	CDC, EPA, FDA, FWS, NIEHS, NOAA, USDA, USGS

Recommendation Categories	2008 HABHRCA Recommendations	Actions Taken	Agencies
		3) Produce clinical therapeutic guidance for the spectrum of illnesses associated with exposure to HAB cells and toxins. Develop case definitions for the spectrum of illnesses resulting from exposure to HAB cells and toxins. Develop communication and outreach programs to make information on HAB poisoning syndromes available to the medical and veterinary community in a timely manner.	
	Improve drinking water monitoring and treatment	Establish guidelines for safe levels of HAB toxins in drinking and recreational water. Monitor source water for HAB cells and toxins, and develop real-time monitoring systems for toxins and cell fragments during water treatment. Develop supplemental water treatment processes for toxins.	EPA, NOAA, USGS

Appendix 4. (Continued)

Recommendation Categories	2008 HABHRCA Recommendations	Actions Taken	Agencies
	Provide HAB forecasts from days to months in advance of HAB events in order to provide managers, industry, and the public with early warning	Develop predictive models that can forecast HABs in a cost-effective and timely manner. Develop food-web based models for the fate and effects of HAB toxins. Develop models of socioeconomic costs of HAB impacts and the cost effectiveness of PCM strategies to support decision-makers.	EPA, NASA, NOAA, NSF
	Mitigate HAB impacts on aquaculture and wild-capture fishery resources		NOAA
	Minimize harvesting bans and closures	Develop effective risk communication to improve public safety. Develop and support strategies to minimize closure areas due to HAB toxin accumulation.	NOAA, FDA
	Modify fishing and processing practices to protect public health	Establish toxin loads in different tissue compartments for commercially harvestable resources and their impacts on the nature of saleable products	FDA, NOAA, USDA

Recommendation Categories	2008 HABHRCA Recommendations	Actions Taken	Agencies
		(e.g., roe-on scallops). Evaluate processing methods that reduce toxin concentrations to below regulatory action levels. Facilitate the application of new sampling and analysis strategies that would provide access to currently restricted resources (e.g., dockside testing protocols for resources on Georges Bank, Geoduck harvest in Alaska).	
	Improve education and outreach	Develop an understanding of public knowledge, attitudes, and perceptions of HABs, and use that understanding or produce effective communication messages for education/outreach programs.	EPA, FDA, NOAA, USDA, USGS
	Community responses to social and economic impacts	Assess and build community resilience to maintain social and economic benefits during HABs.	CDC, NOAA
	Intervene to reduce wildlife mortality		NOAA

Appendix 4. (Continued)

Recommendation Categories	2008 HABHRCA Recommendations	Actions Taken	Agencies
	Understand the Human Dimensions of HAB events in order the assess the efficacy of PCM	Assess the social and economic impacts of HABs Measure the social and economic costs and benefits of PCM strategies to inform societal decision-making. Develop an understanding of public knowledge, attitudes, and perceptions of PCM to produce communication strategies that promote public trust, awareness, and risk-reducing behaviors. Identify and evaluate approaches (e.g., economic incentives, laws, and education) for facilitating changes in human behaviors/attitudes needed to develop and implement PCM strategies. Conduct "institutional analysis" (i.e., research on the nature, strengths, and weaknesses of how people work together) to improve the coordination of researchers,	CDC, EPA, NOAA, USDA, USGS

Recommendation Categories	2008 HABHRCA Recommendations	Actions Taken	Agencies
		decision-makers, and stakeholders involved in PCM research and implementation	
Event Response	Assist during outbreaks of new or unusual HAB events	Provide state and local entities with HAB and toxin testing capabilities and assistance with public outreach during a HAB. Maintain (and deploy as needed) a capability for monitoring and response of HABs that cause unusual events, like seabird, marine mammal mortalities, severely discolored water	CDC, EPA, FDA, FWS, NOAA
Infrastructure	Provide the services and facilities needed to understand and respond to HAB events	Monitoring, training, analytical support during HAB events, data management, toxin standards, tissue bank, algal cultures, HAB forecasting, and information bulletins.	NOAA, EPA, USGS, NIST, NSF, FDA, USDA

APPENDIX 5. ACTIONS TAKEN SINCE 2008 HABHRCA REPORTS – HYPOXIA

Recommendation Categories	2008 HABHRCA Recommendations	Actions Taken	Agencies
Prevention and Mitigation	Nutrient management technology development and improve natural nutrient removal processes	1) Breed plants cultivars that require less phosphorus fertilizers; 2) Minimize unregulated nutrient loading to freshwater systems (NSF Environmental Engineering and Environmental Sustainability Program); 3) Optimize watershed placement of phosphorus and sediment BMPs; 4) Study changes in land use and infrastructure that inform decisions on watershed alterations.	USDA, NSF
	Improve characterization and quantification of hypoxia via surveys	1) Measure dissolved oxygen in the water column and sediments.	BOEM, USGS, NOAA, NSF, EPA, USGS
	Improve characterization and quantification of hypoxia via instrumented observing systems	1) Collect environmental data with instrumented observing systems (e.g., Integrated Ocean Observing System (IOOS) Regional Associations; NSF Ocean Observing Initiative, National Ecological Observatory Network, Division of Ocean Sciences, and Biological Infrastructure).	NSF, NOAA

Recommendation Categories	2008 HABHRCA Recommendations	Actions Taken	Agencies
	Improve characterization and quantification of hypoxia via other advanced technologies (e.g., gliders)	1) Map hypoxia using gliders in several coastal regions, including Gulf of Mexico (NGOMEX) program support); 2) Develop a comprehensive plan to incorporate gliders into a global observation network, and to improve data management, product development, and data/product delivery (IOOS National Glider Network Plan).	NOAA, NSF, ONR
	Monitor nutrient sources and fluxes to coastal and freshwater systems	1) Tracks and reports the sources and quantities of nutrients delivered by streams and groundwater to the Great Lakes, freshwater systems, and the Nation's estuaries; 2) Pioneers new field sensor methods and systems for monitoring and delivering real-time nutrient data needed to address hypoxia issues in freshwaters across the Nation.	NOAA, USDA, USGS
	Quantify nutrient flux to coastal waters by assessing nutrient processes in watersheds	1) Provide model and decision support tools for managing nutrient delivery to evaluate the potential benefits of nutrient management (e.g., USGS SPARROW, USDA APEX and Soil and Water Assessment Tool/Hydrologic Unit Modeling for the United States), NOAA Runoff Risk Advisory Forecast); 2) Quantify the environmental effects of conservation practices and programs and develop the science base for managing the agricultural landscape for environmental quality (Conservation Effects Assessment Project).	EPA, FWS,NASA, NOAA, USDA USGS

Appendix 5. (Continued)

Recommendation Categories	2008 HABHRCA Recommendations	Actions Taken	Agencies
	Assess the biogeochemical pathways that process and recycle nutrients and carbon and ultimately lead to generation and maintenance of hypoxia.	1) Assessment, including statistical, to quantify the relationship between nutrient loading, water column primary production, water column and benthic respiration, and hypoxia; 2) Quantify the relationship between physical processes (local wind strength, wind duration, river discharge volume, water column stratification) and hypoxia magnitude (onset, duration, areal extent, volume); 3) Assess water column and benthic nutrient transformation processes controlling hypoxia.	EPA, NOAA, Naval Research Lab (NRL), NSF, USGS
	Evaluate effectiveness of nutrient reduction strategies in reducing hypoxia, and implement nutrient reduction BMPs	1) Support forecast models to assess the progress of nutrient reduction actions in mitigating hypoxia, and inform managers of the predicted effectiveness of alternative nutrient abatement strategies. Hypoxia forecast models in demonstration phase have been developed for the Gulf of Mexico, Narragansett Bay, Chesapeake Bay, Green Bay, and Lake Erie (NOAA NGOMEX and CHRP); 2) Quantify the environmental effects of conservation practices on spatial and temporal trends in water quality at the small watershed scale and develop the science base for managing agricultural landscapes for environmental quality (CEAP);	EPA, FWS, NASA, NOAA, USACE, USDA, USGS

Recommendation Categories	2008 HABHRCA Recommendations	Actions Taken	Agencies
		3) Estimate the environmental benefits of conservation practices and programs on cropland in large river basins and sub-basins and evaluate conservation treatment needs for remaining soil and water concerns (CEAP); 4) Provide a forum for state water, natural resources, and agricultural agencies and Federal agencies to partner on local, state, and regional nutrient reduction efforts, encouraging an approach that takes into account upstream sources and downstream impacts (Mississippi River/Gulf of Mexico Watershed Nutrient Task Force); 5) Create an integrated framework that supports planning, design, configuration, and delivery of conservation practices within the watershed (Mississippi River Basin/Gulf Hypoxia Initiative); 6) Monitor water chemistry (e.g., National Estuary Program).	
	Conduct surveys to improve characterization of hypoxia impacts	1) Monitor inter-annual estimates of relative abundance for demersal species occurring in the Gulf of Mexico (NOAA SEAMAP); 2) Assess the effects of hypoxia on fish and shrimp distributions in coastal systems (NOAA NGOMEX and CHRP).	NOAA

Appendix 5. (Continued)

Recommendation Categories	2008 HABHRCA Recommendations	Actions Taken	Agencies
	Assess local habitat quality with regard to water quality concerns associated with the presence of hypoxia	Assess the effects of hypoxia on fish and shrimp distributions in coastal systems (NOAA NGOMEX and CHRP).	NOAA
	Quantify the ecosystem impacts of hypoxia	Assess the effects of hypoxia on mortality of managed species and their prey, fecundity (sublethal effects of exposure), habitat and habitat quality, growth (including size), susceptibility to predation, migratory patterns, and bycatch (NOAA NGOMEX and CHRP).	NOAA
	Assess local water quality concerns associated with the presence of hypoxia (e.g., high nitrate in drinking water)	Measure dissolved oxygen and provide warnings to drinking water managers about hypoxic water near the water intake (NOAA Hypoxia Warning System for Cleveland, Ohio).	NOAA
	Improve characterization and quantification of hypoxia ecosystem impacts via improved modeling	Develop ecological models to predict the impacts of hypoxia on living resources at the population, community, and ecosystem levels (NOAA NGOMEX and CHRP). Advances have been used to inform the 2013 reassessment of the Gulf of Mexico Hypoxia Task Force Action Plan.	NOAA

Recommendation Categories	2008 HABHRCA Recommendations	Actions Taken	Agencies
	Quantify nutrient flux to coastal waters and provide decision support tools by modeling source areas, source mechanisms, and trends in nutrient loads	1) Model the linkage between discharge and nutrients from the Mississippi River Basin to Gulf of Mexico hypoxia; 2) Create screening-level TMDL models for nitrogen and phosphorus.	EPA, NOAA, NRL, USGS
	Improve existing and future coordination for data management, monitoring, and research capabilities		EPA, NOAA, USDA, USGS
	Assess economic impacts of hypoxia	Assess the bioeconomic effects of hypoxia on shrimp in the Gulf of Mexico (NGOMEX program) and the Neuse estuary (CHRP program).	NOAA
	Biofuels research		
	Outreach and education	Outreach on hypoxia causes and impacts, nutrient reduction needs, and BMPs	EPA, FWS, NPS, NASA, NOAA, USACE, USDA, USGS

REFERENCES

Aikman, F., D.C. Brady, M.J. Brush, P. Burke, C.F. Cerco, J.J. Fitzpatrick, R. He, G.A. Jacobs, W.M. Kemp, and J.D. Wiggert. 2014. "Modeling Approaches for Scenario Forecasts for Gulf of Mexico Hypoxia." Edited by D.M. Kidwell, A.J. Lewitus, and E. Turner. *White Paper from the Hypoxic Zone Modeling Technical Review*. Silver Spring, Maryland: *NOAA/NOS/NCCOS*. 2014. http://www2.coastalscience.noaa.gov/publications/detail.aspx?resource=WZFA6NrZ8oBLUl7HzHZeLey7Rj DGXfagONIXrpHdEEk.

Allen, A.L., C.W. Brown, A.J. Lewitus, P.A. Sandifer. "The Roles of Emerging Technology and Modeling Techniques in Operational Ecological Forecasting at NOAA." *Marine Technology Society Journal* 49(2/2015): 193-203. DOI: http://dx.doi.org/10.4031/MTSJ.49.2.18.

Anderson, D.M., A.D. Cembella, G. M. Hallegraeff. "Progress in Understanding Harmful Algal Blooms:

Paradigm Shifts and New Technologies for Research, Monitoring, and Management." *Annual Reviews of Marine Science* 2012 (4): 143-176. DOI: 10.1146/annurev-marine-120308-081121.

Backer, L. "Impacts of Florida Red Tides on Coastal Communities." *Harmful Algae* 8(4/2009): 618-622. DOI: 10.1016/j.hal.2008.11.008.

Baden, S.P., L.O. Loo, L. Pihl, and R. Rosenberg. "Effects of Eutrophication on Benthic Communities Including Fish – Swedish West Coast." *Ambio* 19(1990): 113-122. http://www.jstor.org.proxy.lib.umich.edu/stable/4313676.

Baganz, D., G. Staaks, and C. Steinberg. "Impact of the Cyanobacteria Toxin, Microcystin-lr, on Behaviour of Zebrafish, *Danio rerio*." *Water Research* 32(1998): 948-952. DOI: 10.1016/S0043- 1354(97)00207-8.

Bates, S.S., C.J. Bird, A.S.W. de Freitas, R. Foxall, M. Gilgan, L.A. Hanic, G.R. Johnson, A.W. McCulloch, P. Odense, R. Pocklington, M.A. Quilliam, P.G. Sim, J.C. Smith, D.V. Subba Rao, E.C.D. Todd, J.A. Walter, and J.L.C. Wright. "Pennate Diatom *Nitzschia pungens* as the Primary Source of Domoic Acid, a Toxin in Shellfish from Eastern Prince Edward Island, Canada." *Canadian Journal of Fisheries and Aquatic Sciences* 46(1989): 1203-1215. DOI: 10.1139/f89-156.

Beaver, J.R., E.E. Manis, K.A. Loftin, J.L. Graham, A. I. Pollard, R.M. Mitchel. "Land Use Patterns, Ecoregion, and Microcystin Relationships in US Lakes and Reservoirs: A Preliminary Evaluation." *Harmful Algae* 36 (2014): 57-62. DOI: 10.1016/j.hal.2014.03.005.

Boesch, D.F., V.J. Coles, D.G. Kimmel, and W.D. Miller. "Coastal Dead Zones and Global Climate Change: Ramifications of Climate Change for the Chesapeake Bay". In: *Regional Impacts of Climate Change: Four Case Studies in the United States*. Arlington, VA: Pew Center for Global Climate Change. 2007. http://www.c2es.org/docUploads/Regional-Impacts-Chesapeake.pdf.

Boyd, C.E. "Dissolved Oxygen and Other Gases." In Boyd, C.E., *Water Quality: An Introduction 2*. Springer International Publishing. 2015. DOI: 10.1007/978-3-319-17446-4.

Breitburg, D.L. "Effects of Hypoxia, and the Balance between Hypoxia and Enrichment, on Coastal Fishes and Fisheries." *Estuaries* 25(2002): 767-781. http://moritz.botany.ut.ee/~olli/eutrsem/Breitburg02.pdf.

Breitburg, D.L, L. Pihl, and S.E. Kolesar. "Effects of Low Dissolved Oxygen on the Behavior, Ecology and Harvest of Fishes: A Comparison of the Chesapeake and Baltic Systems." In Rabalais, N.N. and R.E. Turner (eds.), *Coastal Hypoxia: Consequences for Living Resources and Ecosystems. Coastal and Estuarine Studies 58*. Washington, DC: American Geophysical Union. 2001. DOI: 10.1029/CE058p0241.

Bresnan, E., K. Davidson, M. Edwards, L. Fernand, R. Gowen, A. Hall, K. Kennington, A. McKinney, S. Milligan, R. Raine, and J. Silke. "Impacts of climate change on harmful algal blooms." *Marine Climate Change Impacts Partnership: Science Review* 4(2013): 1-8. DOI: 10.14465/2013.arc24.236- 243.

Bricker, S., C. Clement, D. Pirhalla, S. Orlando, and D. Farrow. *National Estuarine Eutrophication Assessment. Effects of Nutrient Enrichment in the Nation's Estuaries*. Silver Spring, Maryland: NOAA, National Ocean Service, Special Projects Office and National Centers for Coastal Ocean Science. 1999. http://specialprojects.nos.noaa.gov/projects/cads/ nees/ Eutro_Report.pdf.

Bricker, S.B., D. Lipton, A. Mason, M. Dionne, D. Keeley, C. Krahforst, J. Latimer, and J. Pennock. "Improving Methods and Indicators for Evaluating Coastal Water Eutrophication: A Pilot Study in the Gulf of Maine." *NOAA Technical Report 20*. Silver Spring, Maryland: National Centers for Coastal Ocean Science, Center for Coastal Monitoring and Assessment. 2006. http://ccma.nos.noaa.gov/news/feature/GulfofMaine.html.

Bricker, S., B. Longstaff, W. Dennison , A. Jones, K. Boicourt, C. Wicks, and J. Woerner. "Effects of Nutrient Enrichment in the Nation's Estuaries: A Decade of Change, National Estuarine Eutrophication Assessment

Update." *NOAA Coastal Ocean Program Decision Analysis Series No. 26.* Silver Spring, Maryland: National Centers for Coastal Ocean Science. 2007. http://ccma.nos.noaa.gov/news/feature/Eutroupdate.html.

Bricker, S.B., B. Longstaff, W. Dennison, A. Jones, K. Boicourt, C. Wicks, and J. Woerner. "Effects of Nutrient Enrichment in the Nation's Estuaries: A Decade of Change." *Harmful Algae* 8(2008): 21- 32. DOI:10.1016/j.hal.2008.08.028.

Bridgeman, T.B., D.W. Schloesser, and A.E. Krause. "Recruitment of Hexagenia Mayfly Nymphs in Western Lake Erie Linked to Environmental Variability." *Ecological Applications* 16(2/2006): 601-611. http://www.jstor.org/stable/40061680.

Bridgeman, T.B. and W.A. Penamon. "*Lyngbya wollei* in Western Lake Erie." *Journal of Great Lakes Research* 36(1/2010): 167-171. DOI:10.1016/j.jglr.2009.12.003.

Burns, N.M., D.C. Rockwell, P.E. Bertram, D.M. Dolan, and J.J.H.Ciborowski. "Trends in Temperature, Secchi Depth, and Dissolved Oxygen Depletion Rates in the Central Basin of Lake Erie, 1983-2002." *Journal of Great Lakes Research* 31 (Supplement 2/2005): 35-49. DOI: 10.1016/S0380- 1330(05)70303-8.

Caron, D.A., M.-E. Garneau, E. Seubert, M.D.A. Howard, L. Darjany, A. Schnetzera, I. Cetinic´, G. Filteauc, P. Laurid, B. Jonesa, S. Trusselle. "Harmful Algae and their Potential Impacts on Desalination Operations off Southern California." *Water Research* 44(2/2010): 385-416.DOI:10.1016/j.watres.2009.06.051.

Carpenter, S.R., S.G. Fisher, N.B. Grimm, and J.F. Kitchell. 1992. "Global Change and Freshwater Ecosystems." *Annual Review of Ecology and Systematics* 23(1992): 119-139. DOI: 10.1146/annurev.es.23.110192.001003.

Chaffin, J. D., T. B. Bridgeman, and D. L. Bade. "Nitrogen Constrains the Growth of Late Summer Cyanobacterial Blooms in Lake Erie." *Advances in Microbiology* 3(2013): 16-26. DOI: 10.4236/aim.2013.36A003.

Chakraborty, S. and U. Feudel. "Harmful Algal Blooms: Combining Excitability and Competition." *Theoretical Ecology* 7(2014): 221-237. DOI: 10.1007/s12080-014-0212-1.

Chan, F., J.A. Barth, J. Lubchenco, A. Kirincich, H. Weeks, W.T. Peterson, and B.A. Menge. "Emergence of Anoxia in the California Current Large Marine Ecosystem." *Science* 319(2008): 920. DOI: 10.1126/science.1149016.

Chorus, I. *Current Approaches to Cyanotoxin Risk Assessment, Risk Management and Regulations in Different Countries*. Dessau-Rosslau, Germany: Federal Environment Agency (Umweltbundesamt). 2012. http://www.uba.de/uba-info-medien-e/4390.html.

Committee on Environment and Natural Resources (CENR). *Integrated Assessment of Hypoxia in the Northern Gulf of Mexico*. Washington, DC: National Science and Technology Council Committee on Environment and Natural Resources. 2000. http://oceanservice.noaa.gov/products/hypox_final.pdf.

Committee on Environment and Natural Resources (CENR). *Scientific Assessment of Hypoxia in US Coastal Waters*. Washington, DC: Interagency Working Group on Harmful Algal Blooms, Hypoxia, and Human Health of the Joint Subcommittee on Ocean Science and Technology. 2010. https://www.whitehouse.gov/sites/default/files/microsites/ostp/hypoxia-report.pdf. Last accessed: June 3, 2015.

Conroy, J. D., D.D. Kane, D.M. Dolan, W.J. Edwards, M.N. Charlton, and D.A. Culver "Temporal Trends in Lake Erie Plankton Biomass: Roles of External Phosphorus Loading and Dreissenid Mussels." *Journal of Great Lakes Research* 31(2005): 89-110. DOI:10.1016/S0380-1330(05)70307-5.

Correll, D.L. "The Role of Phosphorus in Eutrophication of Receiving Waters: A Review". *Journal of Environmental Quality* 27(1998): 261-266. http://nature.berkeley.edu/classes/espm120/Website/correll1998.pdf.

Craig, J.K. and S.H. Bosman. "Small Spatial Scale Variation in Fish Assemblage Structure in the Vicinity of Northwestern Gulf of Mexico Hypoxic Zone." *Estuaries and Coasts* 36(2013): 268-285. DOI: 10.1007/s12237-012-9577-9.

Craig, J.K., and L.B. Crowder. "Hypoxia-Induced Habitat Shifts and Energetic Consequences in Atlantic Croaker and Brown Shrimp on the Gulf of Mexico Shelf." *Marine Ecology Progress Series* 294(2005): 79-94. DOI: 10.3354/meps294079.

Craig, J.K., L.B. Crowder, and T.A. Henwood. "Spatial Distribution of Brown Shrimp (*Farfantepenaeus aztecus*) on the Northwestern Gulf of Mexico Shelf: Effects of Abundance and Hypoxia." *Canadian Journal of Fisheries and Aquatic Sciences* 62(2005): 1295-1308. DOI: 10.1139/f05-036.

Craig, J.K. "Aggregation on the Edge: Effects of Hypoxia Avoidance on the Spatial Distribution of Brown Shrimp and Demersal Fishes in the Northern Gulf of Mexico." *Marine Ecological Progress Series* 445 (2012): 75-95. http://www.int-res.com/abstracts/meps/v445/p75-95/.

Davidson, K., R.J. Gowen, P.J. Harrison, L.E. Flemming, P. Hoagland, G. Moschonas. "Anthropogenic Nutrients and Harmful Algae in Coastal Waters." *Journal of Environmental Management.* 146(2014): 206-216. DOI: 10.1016/j.jenvman.2014.07.002.

Davis, T.W, D.L. Berry, G.L. Boyer, and C.J. Gobler. "The Effects of Temperature and Nutrients on the Growth and Dynamics of Toxic and Non-toxic Strains of *Microcystis* During Cyanobacteria Blooms." *Harmful Algae* 8(2009): 715-725. DOI: 10.1016/j.hal.2009.02.004.

Davis, T. W., M. J. Harke, M. A. Marcoval, J. Goleski, C. Orano-Dawson, D. L. Berry, C. J. Gobler. "Effects of Nitrogenous Compounds and Phosphorus on the Growth of Toxic and Non-Toxic Strains of *Microcystis* During Cyanobacterial Blooms." *Aquatic Microbial Ecology* 61(2010): 149-162. DOI: 10.3354/ame01445.

de Mutsert, K., J.H. Cowan, Jr., and C.J. Walters. "Using Ecopath with Ecosim to Explore Nekton Community Response to Freshwater Diversion into a Louisiana Estuary." *Marine and Coastal Fisheries: Dynamics, Management, and Ecosystem Science* 4(1/2012): 104-116. DOI: 10.1080/19425120.2012.672366.

Deeds, J.R., K. Wiles, G.B. Heideman 6th, K.D. White, A. Abraham. "First U.S. Report of Shellfish Harvesting Closures Due to Confirmed Okadaic Acid in Texas Gulf Coast Oysters." *Toxicon* 55(2010): 1138-1146. DOI: 10.1016/j.toxicon.2010.01.003.

Delegrange, A., D. Vincent, L. Courcot, R. Amara. "Testing the Vulnerability of Juvenile Sea Bass (*Dicentrarchus labrax*) Exposed to the Harmful Algal Bloom (HAB) Species *Pseudo-nitzschia delicatissima*." *Aquaculture* 437 (2015): 167-174. DOI: 10.1016/j.aquaculture. 2014.11.023.

Delorme, L. D. "Lake Erie Oxygen: The Prehistoric Record." *Canadian Journal of Fisheries and Aquatic Sciences* 39(1982): 1021–1029. DOI: 10.1139/f82-137.

Diaz, R.J. and R. Rosenberg. "Marine Benthic Hypoxia: A Review of its Ecological Effects and the Behavioural Responses of Benthic Macrofauna." *Oceanography and Marine Biology an Annual Review* 33(1995): 245-303. http://www.researchgate.net/profile/Robert_Diaz5/publication/236628341_Marine_benthic_hypoxia_A_review_of_its_ecological_effects_and_the_behavioural_responses_of_benthic_macrofauna/links/02e7e526a7c717396d000000.pdf.

Diaz, R.J. and R. Rosenberg. "Spreading Dead Zones and Consequences for Marine Ecosystems." *Science* 321(2008): 926-929. DOI: 10.1126/science.1156401.

Diaz, R.J. and R. Rosenberg. "Introduction to Environmental and Economic Consequences of Hypoxia." *International Journal of Water Resources Development* 27(2011): 71-82. DOI: 10.1080/07900627.2010.531379.

Dortch, Q., D.M. Anderson, D.L. Ayres, and P.M. Glibert. *Harmful Algal Bloom Research, Development, Demonstration and Technology Transfer: A National Workshop Report.* Woods Hole, Massachusetts: Woods Hole Oceanographic Institute. 2008. http://www.whoi.edu/fileserver.do?id=43464&pt=10&p=19132.

Dyson, K., D. D. Huppert. "Regional Economic Impacts of Razor Clam Beach Closures Due to Harmful Algal Blooms (HABs) on the Pacific Coast of Washington." *Harmful Algae* 9(2010): 264-271. doi:10.1016/j.hal. 2009.11.003.

Dziuban, E.J., J.L. Liang, G.F. Craun, V. Hill, P.A. Yu, J. Painter, M.R. Moore, R.L. Calderon, S.L. Roy, and M.J. Beach. "Surveillance for Waterborne Disease and Outbreaks Associated with Recreational Water – United States, 2003-2004." *Morbidity and Mortality Weekly Report Surveillance Summaries* 55(2006): 1-30. http://www.cdc.gov/mmwr/PDF/ss/ss5512.pdf.

Ecological Society of America. "The Role Of Atmospheric Deposition In The Gulf Of Mexico Hypoxic Zone: Workshop Report." http://www.esa.org/esa/science/reports/hypoxia/. Last accessed: June 5, 2015.

Etheridge, S. "Paralytic Shellfish Poisoning: Seafood Safety and Human Health Perspectives." *Toxicon* 56 (2010/2): 108-122. DOI: 10.1016/j.toxicon.2009.12.013.

Francesconi, W., Smith, D.R., Heathman, G.C., Wang, X., Williams, C. "Monitoring and APEX modeling of no-till and reduced-till in tile drained agricultural landscapes for water quality." Transactions of the ASABE 57(3/2014):777-789. http://www.researchgate.net/profile/Wendy_Francesconi/publication/264120739_MONITORING_AND_APEX_MODELING_OF_NO-TILL_AND_REDUCED-TILL_IN_TILE-DRAINED_AGRICULTURAL_LANDSCAPES_FOR_WATER_QUALITY/links/53cec9da0cf2f7e53cf7d1dc. pdf.

Fujiki, H. and M. Suganuma. "Tumor Promoters-Microcystin-LR, Nodularin and TNF-α and Human Cancer Development." *Anti-Cancer Agents in Medicinal Chemistry (Formerly Current Medicinal ChemistryAnti-Cancer Agents)* 11.1(2011): 4-18. DOI: 10.2174/187152011794941163.

Genitsaris, S., K.A. Kormas, M. Moustaka-Gouni. "Airborne Algae and Cyanobacteria: Occurrence and Related Health Effects." *Frontiers in Bioscience* 3(2011): 772-787. http://kkormas.users.uth.gr/genitsaris FB.pdf.

Giannuzzi, L., D. Sedan, R. Echenique, and D. Andrinolo. "An Acute Case of Intoxication with Cyanobacteria and Cyanotoxins in Recreational Water in Salto Grande Dam, Argentina." *Marine Drugs* 9(2011): 2164-2175. DOI: 10.3390/md9112164.

Gingold, D., M.J. Strickland, and J.J. Hess. "Ciguatera Fish Poisoning and Climate Change: Analysis of National Poison Center Data in the United States, 2001-2011." *Environmental Health Perspectives* 122(2014): 580-586. DOI: http://dx.doi.org/10.1289/ehp.1307196.

Graham, J. L., K. A. Loftin, and N. Kamman. "Monitoring Recreational Freshwaters." *Lakeline* (2009): 18- 24. http://ks.water.usgs.gov/static_pages/studies/water_quality/cyanobacteria/LLsummergraham2.pdf.

Great Lakes Water Quality Protocol of 2012. Signed September 7, 2012. Entered into force February 12, 2013. https://www.ec.gc.ca/grandslacs-greatlakes/A1C62826-72BE-40DB-A545-65AD6FCEAE92/1094_Canada-USA%20GLWQA%20_e.pdf. Last accessed: August 17, 2015.

Greenfield, D.I., A. Duquette, A. Goodson, C.J. Keppler, S.H. Williams, L.M. Brock, K.D. Stackley, D. White, and S. Wilde. "The Effects of Three Chemical Algaecides on Cell Numbers and Toxin Content of the Cyanobacteria *Microcystis aeruginosa* and *Anabaenopsis* sp." *Environmental Management* 54(2014): 1110-1120. DOI: 10.1007/s00267-014-0339-2.

Griffis, R., and J. Howard, ed. 2013. *Oceans and Marine Resources in a Changing Climate: Technical Input to the 2013 National Climate Assessment*. Washington, DC: Island Press, 249 pp.

Grimes, D.J., M.T. Hamann, J.M. Lotz, T.I. McLean, T. McIlwain, and C.H. Price. "Oceans and Human Health, Social and Economic Impacts." In: R.A. Meyers (ed.), *Encyclopedia of Sustainability Science and Technology*. New York: Springer 2012. DOI: 10/1007/978-1-4419-0851-3_834.

Gulland, F.M.D. and A.J. Hall. "Is Marine Mammal Health Deteriorating? Trends in the Global Reporting of Marine Mammal Disease." *EcoHealth* 4(2007): 135-150. DOI: 10.1007/s10393-007-0097-1.

Hamidi, S.A., H.R. Bravo, and J.V. Klump. "Evidence of Multiple Physical Drivers on the Circulation and Thermal Regime in the Green Bay of Lake Michigan." *World Environmental and Water Resources Congress 2013:*

Showcasing the Future. (2013): 1719-1726. DOI: 10.1061/9780784412947.169.

HARRNESS. *Harmful Algal Research and Response: A National Environmental Science Strategy 2005- 2015.* J.S. Ramsdell, D.M. Anderson, and P.M. Glibert (eds.) Washington, DC: Ecological Society of America. 2005. http://www.esa.org/HARRNESS/harrnessReport10032005.pdf.

Hazen, E. L., J. K. Craig, C. P. Good, and L. B. Crowder. "Vertical Distribution of Fish Biomass in Hypoxic Waters on the Gulf of Mexico Shelf." *Marine Ecological Progress Series* 375(2009): 195-207. DOI: 10.3354/meps07791.

Hilborn, E.D., V.A. Roberts, L. Backer, E. DeConno, J.S. Egan, J.B. Hyde, D.C. Nicholas, E.J. Wiegert, L.M. Billing, M. DiOrio, M.C. Mohr, F.J. Hardy, T.J. Wade, J.S. Yoder, and M.C. Hlavsa. "Algal Bloom-Associated Disease Outbreaks Among Users of Freshwater Lakes – United States, 2009-2010." *Morbidity and Mortality Weekly Report (MMWR)* 63(2014): 11-15. http://www.cdc.gov/mmwr/preview/mmwrhtml/mm6301a3.htm.

Hilborn, E.D., V.R. Beasley. "One Health and Cyanobacteria in Freshwater Systems: Animal Illnesses and Deaths Are Sentinel Events for Human Health Risks." *Toxins* 7(2015): 1374-1395. DOI: 10.3390/toxins7041374.

Hoagland, P. and S. Scatasta. "The Economic Effects of Harmful Algal Blooms." In: E. Graneli and J.T. Turner (eds.), *Ecology of Harmful Algae.* Berlin, Germany: Springer. 2006. DOI: 10/1007/978-3-540- 32210-8_30.

Hollister, J.W., W.B Milstead, B.J. Kreakie. "Modeling Lake Trophic State: A Random Forest Approach." (In revision.)

Horst, G.P., O. Sarnelle, J.D. White, S.K. Hamilton, R.B. Kaul, and J.D. Bressie. "Nitrogen Availability Increases the Toxin Quota of Harmful Cyanobacterium, *Microcystis aeruginosa*." *Water Research* 54(2014): 188-198. DOI: 10.1016/j.watres.2014.01.063.

Howden, S.D, R.A. Arnone, J. Brodersen, S.F. DiMarco, L.K. Dixon, H.E. Garcia, M.K. Howard, A.E. Jochens, S.E. Ladner, C.E. Lembke, A.P. Leonardi, A. Quaid, and N.N. Rabalais. *Glider Implementation Plan for Hypoxia Monitoring in the Gulf of Mexico.* Edited by A.J. Lewitus, S.D. Howden, and D.M. Kidwell. National Oceanic and Atmospheric Administration. 2014. http://coastalscience.noaa.gov/news/wp-content/uploads/2014/05/Glider-Implementation-Planfor-Hypoxia-Monitoring-in-the-Gulf-of-Mexico.pdf.

Hudnell, H. K., Q. Dortch, and H. Zenick. "An Overview of the Interagency, International Symposium on Cyanobacterial Harmful Algal Blooms

(ISOC-HAB): Advancing the Scientific Understanding of Freshwater Harmful Algal Blooms." *Advances in Experimental Medicine and Biology* 619(2008): 1- 16. DOI: 10.1007/978-0-387-75865-7_1.

Institute of Medicine (US) Committee on Evaluation of the Safety of Fishery Products. "Naturally Occurring Fish and Shellfish Poisons." Ahmed FE, editor. *Seafood Safety*. Washington (DC): National Academies Press (US). 1991(4). http://www.ncbi.nlm.nih.gov/books/NBK235731/. Last accessed: May 12, 2015.

Jewett, E.B., C.B. Lopez, Q. Dortch, S.M. Etheridge. "National Assessment of Efforts to Predict and Respond to Harmful Algal Blooms in US Waters. Interim Report." Washington, DC: Interagency Working Group on Harmful Algal Blooms, Hypoxia, and Human Health of the Joint Subcommittee on Ocean Science and Technology. 2007.

Jewett, E.B., C.B. Lopez, Q. Dortch, S.M. Etheridge, and L.C. Backer. *Harmful Algal Bloom Management and Response: Assessment and Plan.* Washington, DC: Interagency Working Group on Harmful Algal Blooms, Hypoxia and Human Health of the Joint Subcommittee on Ocean Science and Technology 2008. http://www.cop.noaa.gov/stressors/extremeevents/hab/habhrca/HABMngmt_resp_9_08.pdf.

Justic´, D., V.J. Bierman, Jr., D. Scavia, and R.D. Hetland. "Forecasting Gulf's Hypoxia: The Next 50 Years?" *Estuaries and Coasts* 30(5/2007): 791-801. http://www.jstor.org/stable/27654716.

Karlson, K., R. Rosenberg, and E. Bonsdorff. "Temporal and Spatial Large-Scale Effects of Eutrophication and Oxygen Deficiency on Benthic Fauna in Scandinavian and Baltic Waters: A Review." *Oceanography and Marine Biology* 40(2002): 427-489. DOI: 10.1.1.177.8144.

Kemp, W.M., W.R. Boynton, J.E. Adolf, D.F. Boesch, W.C. Boicourt, G. Brush, J.C. Cornwell, T. R. Fisher, P.M. Gilbert, J.D. Hagy, L.W. Harding, E.D. Houde, D.G. Kimmel, W.D. Miller, R.I. E. Newell, M.R. Roman, E.M. Smith, and J.C. Stevenson. "Eutrophication of Chesapeake Bay: Historical Trends and Ecological Interactions." *Marine Ecology Progress Series* 303(2005): 1-29. DOI: 10.3354/meps303001.

Kraus, R.T., C.T. Knight, T.M. Farmer, A.M. Gorman, P.D. Collingsworth, G.J. Warren, P.M. Kocovsky, and J.D. Conroy. "Dynamic Hypoxic Zones in Lake Erie Compress Fish Habitat, Altering Vulnerability to Fishing Gears." *Canadian Journal of Fisheries and Aquatic Sciences* 72(2015): 1-10. DOI: 10.1139/cjfas-2014-0517.

Kudela, R.M., E. Berdalet, S. Bernard, M. Burford, L. Fernand, S. Lu, S. Roy, P. Tester, G. Usup, R. Magnien, D.M. Anderson, A. Cembella, M.

Chinain, G. Hallegraeff, B. Reguera, A. Zingone, H. Enevoldsen, E. Urban. *A Scientific Summary for Policy Makers*. IOC/UNESCO, Paris (IOC/INF-1320). http://unesdoc.unesco.org/images/0023/002334/233419e.pdf. Last accessed: August 10, 2015.

Lapointe, B.E., P.J. Barile, W.R. Matzie. "Anthropogenic Nutrient Enrichment of Seagrass and Coral Reef Communities in the Lower Florida Keys: Discrimination of Local Versus Regional Nitrogen Sources." *Journal of Experimental Marine Biology and Ecology,* 308(1/2004). 308:23-58.

Lehane, L. and R.J. Lewis. "Ciguatera: Recent Advances but the Risk Remains." *International Journal of Food Microbiology* 61(2000): 91-125. DOI: 10.1016/S0168-1605(00)00382-2.

Lewitus, A., R.A. Horner, D.A. Caron, E. Garcia-Mendoza, B.M. Hickey, M. Hunter, D.D. Huppert, R.M. Kudela, G.W. Langlois, J.L. Largier, E.J. Lessard, R. RaLonde, J.E. Jack Rensel, P.G. Strutton, V.L. Trainer, J.F. Tweddle. "Harmful Algal Blooms Along the North American West Coast Region: History, Trends, Causes, and Impacts." *Publications, Agencies and Staff of the US Department of Commerce* (499/2012). http://digitalcommons.unl.edu/usdeptcommercepub/499.

Lipton, D. W. and R. Hicks. "Linking Water Quality Improvements to Recreational Fishing Values." In: *Evaluating the Benefits of Recreational Fisheries. Fisheries Centre Research Reports* 7(2/1999). University of British Columbia.

Lipton D., and R. Hicks. "The Cost of Stress: Low Dissolved Oxygen and Economic Benefits of Recreational Striped Bass (*Morone saxatilis*) Fishing in the Patuxent River." *Estuaries* 26(2003): 310-315. http://www.jstor.org/stable/1353127.

Lopez, C. B., E. B. Jewett, Q. Dortch, B. T. Walton, and H. K. Hudnell. Scientific Assessment of Freshwater Harmful Algal Blooms. Washington, D.C. : Interagency Working Group on Harmful Algal Blooms, Hypoxia, and Human Health of the Joint Subcommittee on Ocean Science and Technology. 2008a. https://www.whitehouse.gov/sites/default/files/microsites/ostp/frshh2o0708.pdf.

Lopez, C.B., Q. Dortch, E.B. Jewett, and D. Garrison. Scientific Assessment of Marine Harmful Algal Blooms. Washington, DC: Interagency Working Group on Harmful Algal Blooms, Hypoxia, and Human Health of the Joint Subcommittee on Ocean Science and Technology. 2008b. https://www.whitehouse.gov/sites/default/files/microsites/ostp/jsost-assess_12-08.pdf.

Lunetta, R.S., B.A. Schaeffer, R.P. Stumpf, D. Keith, S.A. Jacobs, M.S. Murphy. "Evaluation of cyanobacteria cell count detection derived from MERIS imagery across the eastern USA." *Remote Sensing of Environment* 157(2015): 24-34. DOI: 10.1016/j.rse.2014.06.008.

Lürling, M. and I. Roessink. "On the Way to Cyanobacterial Blooms: Impact of the Herbicide Metribuzin on the Competition between a Green Alga (Scenedesmus) and a Cyanobacterium (Microcystis)." *Chemosphere* 65(4/2006): 618-626. DOI:10.1016/j.chemosphere.2006.01.073.

Marinho, M.M., Souza, M.B.G., Lurling, M. "Light and phosphorous competition between *Cylindrospermopsis raciborskii* and *Microcystis aeruginosa* is strain dependent." *Microbial Ecology* 66(2013): 479-488. DOI: 10.1007/s00248-013-0232-1.

Maryland Department of Natural Resources. Eyes on the Bay. 2015. http://mddnr.chesapeakebay.net/eyesonthebay/.

McCoy, C., R. Viso, R.N. Peterson, S. Libes, B. Lewis, J. Ledoux, G. Voulgaris, E. Smith, D. Sanger. "Radon as an Indicator of Limited Cross-Shelf Mixing of Submarine Groundwater Discharge Along an Open Ocean Beach in the South Atlantic Bight During Observed Hypoxia." *Continental Shelf Research* 31(2011): 1306-1317. DOI: 10.1016/j.csr.2011.05.009.

McGillicuddy, D.J., Jr., D.M. Anderson, D.R. Lynch, and D.W. Townsend. "Mechanisms Regulating Large-Scale Seasonal Fluctuations in *Alexandrium fundyense* Populations in the Gulf of Maine: Results from a Physical-Biological Model." *Deep-Sea Research* 52(2005): 2698-2714. DOI:10.1016/j.dsr2.2005.06.021.

Mee, L. "An Achievable goal in a new Europe?" In: *European Conference on Marine Science and Ocean Technology*. Luxembourg: Office for Official Publications of the European Communities. 2007. http://www.eurocean conferences.eu/attachments/eurocean2004_report_declaration.pdf.

Melillo, J.M., T.C. Richmond, G.W. Yohe, Eds., 2014. *Climate Change Impacts in the United States: The Third National Climate Assessment*. U.S. Global Change Research Program, 841 pp. DOI: 10/7930/J0Z31WJ2.

Michalak, A.M., E.J. Anderson, D. Beletski, S. Boland, N.S. Bosch, T.B. Bridgeman, J.D. Chaffin, K. Cho, R. Confesor, I. Daloglu, J.V. DePinto, M.A. Evans, G.L. Fahnenstiel, L. He, J.C. Ho, L. Jenkins, T.H. Johengen, K.C. Kuo, E. LaPorte, X. Liu, M.R. McWilliams, M.R. Moore, D.J. Posselt, R.P. Richards, D. Scavia, A.L. Steiner, E. Verhamme, D.M. Wright, and M.A. Zagorski. "Record-Setting Algal Bloom in Lake Erie Caused by Agricultural and Meteorological Trends Consistent with

Expected Future Conditions." *Proceedings of the National Academy of Sciences* 110.16 (2013): 6448-6452. DOI: 10.1073/pnas.1216006110.

Moore, S.K., N.J. Mantua, and E.P. Salathé, Jr. "Past Trends and Future Scenarios for Environmental Conditions Favoring the Accumulation of Paralytic Shellfish Toxins in Puget Sound Shellfish." *Harmful Algae* 10 (2011): 521-529. DOI: 10.1016/j.hal.2011.04.004.

Moore, W. S. "The Effect of Submarine Groundwater Discharge on the Ocean." *Marine Science* 2 (2010): 59-88. DOI: 10.1146/annurev-marine-120308-081019.

Munday, R. "Toxicology of Seafood Toxins: A Critical Review." In: Botana, L.M. *Seafood and Freshwater Toxins: Pharmacology, Physiology, and Detection*. 2014. ISBN-13: 978-1466505148.

Murphy, T.P., A. Lawson, M. Kumagai, and J. Babin. "Review of Emerging Issues in Sediment Treatment." *Aquatic Ecosystem Health and Management* 2(1999): 419-434. DOI: 10.1080/14634989908656980.

Murphy, C.A., K.A. Rose, M.S. Rahman, and P. Thomas. "Testing and Applying a Fish Vitellogenesis Model to Evaluate Laboratory and Field Biomarkers of Endocrine Disruption in Atlantic Croaker (*Micropogonias undulates*) Exposed to Hypoxia." *Environmental Toxicology and Chemistry* 28(6/2009): 1288-1303. DOI: 10.1897/08-304.1.

Najjar, R.G., H.A. Walker, P.J. Anderson, E.J. Barron, R.J. Bord, J.R. Gibson, V.S. Kennedy, C.G. Knight, J.P. Megonigal, R.E. O'Connor, C.D. Polsky, N.P. Psuty, B.A. Richards, L.G. Sorenson, E.M. Steele, and R.S. Swanson. "The potential Impacts of Climate Change on the Mid-Atlantic Coastal Region." *Climate Research* 14(2000): 219-233. http://www.cara.psu.edu/about/publications/Najjar_et_al_2000.pdf.

National Oceanic and Atmospheric Administration (NOAA). "Interim Report on the Bottlenose Dolphin (Tursiops truncatus) Unusual Mortality Event Along the Panhandle of Florida, March-April 2004." 2004. http://www.nmfs.noaa.gov/pr/pdfs/health/ume bottlenose 2004.pdf. Last accessed: January 22, 2016.

National Oceanic and Atmospheric Administration (NOAA). *COAST's National Status and Trends, NS&T Program Download Page.* 2012. http://ccma.nos.noaa.gov/about/coast/nsandt/download.aspx. Last accessed: April 29, 2015.

National Oceanic and Atmospheric Administration (NOAA). "2015 Gulf of Mexico Dead Zone 'Above Average'". Press release. August 4, 2015. http://www.noaanews.noaa.gov/stories2015/080415- gulf-of-mexico-dead-zone-above-average.html.

National Research Council, Committee on Human and Environmental Exposure Science in the 21st Century; Board on Environmental Studies and Toxicology; Division on Earth and Life Studies. *Exposure Science in the 21st Century: A Vision and a Strategy.* Washington, DC: The National Academies Press, 2012. http://www.ncbi.nlm.nih.gov/books/NBK206816/.

O'Neil, J.M., T.W., Davis, M.A. Burford, and C.J. Gobler. "The Rise of Harmful Cyanobacteria Blooms: The Potential Roles of Eutrophication and Climate Change." *Harmful Algae* 14(2012): 313-334. DOI: 10.1016/j.hal.2011.10.027.

Paerl, H.W., and J. Huisman. "Blooms Like it Hot." *Science* 320(2008): 57-58. DOI: 10.1126/science.1155398.

Paerl, H.W. and V.J. Paul. "Climate Change: Links to Global Expansion of Harmful Cyanobacteria." *Water Research.* 46(2012): 1349-1363. DOI: doi:10.1016/j.watres.2011.08.002.

Paerl, H.W. and T.G. Otten. "Harmful Cyanobacterial Blooms: Causes, Consequences, and Controls." *Microbial Ecology* 46(5/2013): 1349-1363. DOI: 10.1007/s00248-012-0159-y.

Pellerin, B.A., B.A. Bergamaschi, R.J. Gilliom, C.G. Crawford, J. Saraceno, C.P. Frederick, B.D. Downing, and J.C. Murphy. "Mississippi River Nitrate Loads from High Frequency Sensor Measurements and Regression-Based Load Estimation." *Environmental Science & Technology* 48(2014): 12612−12619. DOI: 10.1021/es504029c.

Peterson, H.G., C. Boutin, K.E. Freemark, and P.A. Martin. "Toxicity of hexazinone and diquat to green algae, diatoms, cyanobacteria and duckweed." *Aquatic Toxicology* 39(1997): 111-134. DOI:10.1016/S0166-445X(97)00022-2.

Pierson, J.J., M.R. Roman, D.G. Kimmel, W.C. Boicourt, and X.Zhang. "Quantifying Changes in Vertical Distribution of Mezoplankton in Response to Hypoxic Bottom Waters." *Journal of Experimental Marine Biology and Ecology* 381(2009): S74-S79. DOI:10.1016/j.jembe.2009.07.013.

Rabalais, N.N. and R.E. Turner. "Coastal Hypoxia: Consequences for Living Resources and Ecosystems." *Coastal Estuarine Studies* 58. Washington, DC: American Geophysical Union. 2001. DOI: 10.1029/CE058.

Rabalais, N.N., R.E. Turner, and W.J. Wiseman, Jr. "Hypoxia in the Gulf of Mexico." *Journal of Environmental Quality* 30(2001):320-329. DOI:10.2134/jeq2001.302320x.

Rabalais, N.N., R.E. Turner, B.K. Gupta, D.F. Boesch, P. Chapman, and M.C. Murrell. "Hypoxia in the
Northern Gulf of Mexico: Does the Science Support the Plan to Reduce, Mitigate, and Control Hypoxia?" *Coastal and Estuarine Research Federation* 30(5/2007): 753-772.

Rabotyagov, S. S., C.L. Kling, P.W. Gassman, N.N. Rabalais, and R.E. Turner. "The Economics of Dead Zones: Causes, Impacts, Policy Changes, and a Model of the Gulf of Mexico Hypoxic Zone." *Review of Environmental Economics and Policy* 8(1/2014) 58-79. DOI: 10.1093/reep/ret024.

Rahman, M.S. and P. Thomas. "Molecular Cloning, Characterization and Expression of Two Tryptophan Hydroxylase (TPH-1 and TPH-2) Genes in the Hypothalamus of Atlantic Croaker: Down-Regulation after Chronic Exposure to Hypoxia." *Neuroscience* 158(2009): 751-765. DOI:10.1016/j.neuroscience.2008.10.029.

Rahman, M.S., and P. Thomas. "Characterization of Three IGFBP mRNAs in Atlantic Croaker and their Regulation During Hypoxic Stress: Potential Mechanisms of their Upregulation by Hypoxia." *American Journal of Physiology Endocrinology and Metabolism* 301(4/2011): E637-648. DOI: 10.1152/ajpendo.00168.2011.

Rahman, M.S., I.A. Khan, and P. Thomas. "Tryptophan Hydroxylase: A Target for Neuroendocrine Disruption." *Journal of Toxicology and Environmental Health, Part B: Critical Reviews* 14(5-7/2011):
473-494. DOI: 10.1080/10937404.2011.578563.

Rahman, M.S. and P. Thomas. "Effects of Hypoxia Exposure on Hepatic Cytochrome P450 (CYP1A)
Expression in Atlantic Croaker: Molecular Mechanisms of CYP1A Down-Regulation." *PLOS One* (7/2012). DOI: 10.1371/journal.pone.0040825.

Ramsdell, J.S. and F.M. Gulland. "Domic Acid Epileptic Disease." *Marine Drugs* 12(3/2014): 1185-1207. DOI:10.3390/md12031185.

Rief, M.K. "Remote Sensing for Inland Water Quality Monitoring: A US Army Corps of Engineers perspective." *US Army Corps of Engineers ERDC/EL TR-11-13.* Kiln, Mississippi: Environmental Laboratory, Engineer Research and Development Center. 2011. http://acwc.sdp.sirsi.net/client/en_US/search/asset/1005263;jsessionid=A3303A1D6663B0C3E80E E25EAE424A3B.enterprise-15000.

Riegl, B.M., A.W. Bruckner, K. Samimi-Namin, S.J. Purkis. "Diseases, Harmful Algae Blooms (HABs) and their Effects on Gulf Coral Populations and Communities." *Coral Reefs of the Gulf, Coral Reefs of the World* 3(2012): 107-125. Leiden, The Netherlands: Springer Netherlands. 2012. DOI: 10.1007/978-94-007-3008-3_7.

Ritter, C. and P.A. Montagna. "Seasonal Hypoxia and Models of Benthic Response in a Texas Bay." *Estuaries* 22(1/1999): 7-20. http://www.jstor.org/stable/1352922.

Roman, M.R., J.J. Pierson, D. G. Kimmel, W.C. Boicourt, and X. Zhang. "Impacts of Hypoxia on Zooplankton Spatial Distributions in the Northern Gulf of Mexico." *Estuaries and Coasts* 35(2012): 1261-1269. DOI 10.1007/s12237-012-9531-x.

Rosenblatt, A.E., M.R. Heithaus, M.E. Mather, P. Matich, J. Nifong, W.J. Ripple, and B. Silliman. "The roles of Large Top Predators in Coastal Ecosystems: New Insights from Long Term Ecological Research." *Oceanography* 26(2013): 156-167. DOI: 10.5670/oceanog.2013.59.

Sagasti, A., L.C. Schaffner, and J.E. Duffy. "Effects of Periodic Hypoxia on Mortality, Feeding and Predation in an Estuarine Epifaunal Community." *Journal of Experimental Marine Biology and Ecology* 258(2001): 257-283. DOI: 10.1016/S0022-0981(01)00220-9.

Sanford, W.E. and J.P. Pope. "Quantifying Groundwater's Role in Delaying Improvements to Chesapeake Bay Water Quality." *Environmental Science and Technology* 47 (2013/23): . DOI: 10.1021/es401334k.

Scavia, D. and K.A. Donnelly. "Reassessing Hypoxia Forecasts for the Gulf of Mexico." *Environmental Science and Technology* 41(2007): 8111-8117. DOI: 10.1021/es0714235.

Scavia, D., D. Justić, and V.J. Bierman, Jr. "Reducing Hypoxia in the Gulf of Mexico: Advice from Three Models." *Coastal and Estuarine Research Federation* 27(3/2004): 419-425. DOI: 10.1007/BF02803534.

Scavia, D., J.D. Allan, K.K. Arend, S. Bartel, D. Beletsky, N.S. Bosch, S.B. Brandt, R.D. Briland, I. Daloğlu, J.V. DePinto, D.M. Dolan , M.A. Evans, T.M. Farmer, D. Goto, H. Han, T.O. Höök, R. Knight , S.A. Ludsin, D. Mason, A.M. Michalak, R.P. Richards, J.J. Roberts, D. K. Rucinski, E. Rutherford, D.J. Schwab, T.M. Sesterhenn, H. Zhang, Y. Zhou. "Assessing and Addressing the Re-eutrophication of Lake Erie: Central Basin Hypoxia." *Journal of Great Lakes Research* 40.2 (2014): 226-246. DOI:10.1016/j.jglr.2014.02.004.

Schottler, S.P., J. Ulrich, P. Belmont, R. Moore, J.W. Lauer, D.R. Engstrom, and J. E. Almendinger. "Twentieth Century Agricultural Drainage Creates More Erosive Rivers." *Hydrological Processes* 28(2014): 1951-1961. DOI: 10.1002/hyp.9738.

Seltenrich, N. "Keeping Tabs on HABs: New Tools for Detecting, Monitoring, and Preventing Harmful Algal Blooms." *Environmental Health Perspective* 122(2014): A206-A213. DOI: 10.1289/ehp.122- A206.

Shumway, S.E. "A Review of the Effects of Algal Blooms on Shellfish and Aquaculture." *Journal of the World Aquaculture Society* 21(2/1990): 65-104. DOI: 10.1111/j.1749-7345.1990.tb00529.x.

Simeone, C.A., F.M.D. Gulland, T. Norris, T.K. Rowles. "A Systematic Review of Changes in Marine Mammal Health in North America, 1972-2012: The Need for a Novel Integrated Approach." *PLoSOne* 10(11): e0142105. doi: 10.1371/journal.pone.0142105.

Smith, D.R., King, K.W., Johnson, L., Francesconi, W., Richards, P., Baker, D., Sharpley, A.N. "Surface runoff and tile drainage transport of phosphorus in the Midwestern United States." *Journal of Environmental Quality* 44 (2/2014): 495-502. DOI:10.2134/jeq2014.04.0176.

Smith, D.R., K.W. King, M.R. Williams. "What is Causing the Harmful Algal Blooms in Lake Erie?" *Journal of Soil and Water Conservation* 70(2015): 27A-29A. DOI: 10.2489/jswc.70.2.27A.

Smith, M.D., F. Asche, L.S. Bennear, A. Oglend. "Spatial-Dynamics of Hypoxia and Fisheries: The Case of Gulf of Mexico Brown Shrimp." *Marine Resources Economics* 29(2014): 111-131. DOI: http://dx.doi.org/10.1086/676826.

Solomon, S., D. Qin, M. Manning, Z. Chen, M. Marquis, K. B. Averyt, M. Tignor, and H.L. Miller. "Climate Change 2007: The physical science basis," Contribution of Working Group I to the Fourth Assessment Report of the Intergovernmental Panel on Climate Change. Cambridge University Press, New York. 2007.

Stow, C.A. and T. Hook, Eds., with D. Beletsky, R. Beletsky, A. Burtner, J.H. Bredin, J. Cavaletto, Y. Cha, C. De Marchi, J.V. DePinto, J. Dyble, D.Fanslow, S. Francoeur, D. Gossiaux, N. Hawley, C. He, T.H. Johengen, D.R. Kashian, M.D. Kaplowitz, W. Keiper, P.J. Lavrentyev, F. Lupi, D.F. Millie, N. Morehead, T.F. Nalepa, T.J. Newcomb, D. Palladino, S.D. Peacor, S.A. Pothoven, T. Redder, M. Selzer, H.A. Vanderploeg, E. Verhamme, K.Winslow. 2013. "Saginaw Bay Management Report." *NOAA Technical Memorandum GLERL* 160. Ann Arbor, Michigan: NOAA, Great Lakes Environmental Research Laboratory. 2013. http://www.glerl.noaa.gov/ftp/publications/tech_reports/glerl-160/tm-160.pdf. Last accessed: May 15, 2015.

Stumpf, R.P., M.C. Tomlinson, J.A. Calkins, B. Kirkpatrick, K. Fisher, K. Nierenberg, R. Currier, and T.T. Wynne. "Skill Assessment for an

Operational Algal Bloom Forecast System." *Journal of Marine Systems* 76(1-2/2009): 151-161. DOI: 10.1016/j.jmarsys.2008.05.016.

Svirčev, Z., S. Krstic, M. Miladinov-Mikov, V. Baltic, M. Vidovic. "Freshwater Cyanobacterial Blooms and Primary Liver Cancer Epidemiological Studies in Serbia." *Journal of Environmental Science and Health, Part C: Environmental Carcinogenesis and Ecotoxicology Reviews* 27(1/2009): 36-55. DOI: 10.1080/10590500802668016.

Taylor, M., L. McIntyre, M. Ritson, J. Stone, R. Bronson, O. Bitzikos, W. Rourke, E. Galanis, Outbreak Investigation Team. "Outbreak of Diarrhetic Shellfish Poisoning Associated with Mussels, British Columbia, Canada." *Marine Drugs* 11(2013): 1669-1676. DOI: 10.3390/md11051669.

Thomas, P., and M.S. Rahman. "Chronic Hypoxia Impairs Gamete Maturation in Atlantic Croaker Induced by Progestins through Nongenomic Mechanisms Resulting in Reduced Reproductive Success." *Environmental Science & Technology* 43(11/2009a): 4175-4180. DOI: 10.1021/es9000399.

Thomas, P., and S.M. Rahman. "Biomarkers of Hypoxia Exposure and Reproductive Function in Atlantic Croaker: A Review with Some Preliminary Findings from the Northern Gulf of Mexico Hypoxic Zone." *Journal of Experimental Marine Biology and Ecology* 381 (Suppl./2009b): S38-S50. doi:10.1016/j.jembe.2009.07.008

Thomas, P. and M.S. Rahman. "Region-Wide Impairment of Atlantic Croaker Testicular Development and Sperm Production in the Northern Gulf of Mexico Hypoxic Dead Zone." *Marine Environmental Research* 69(2010): 559-562. DOI:10.1016/j.marenvres.2009.10.017

Thomas, R. and M.S. Rahman. "Extensive Reproductive Disruption, Ovarian Masculinization and Aromatase Suppression in Atlantic Croaker in the Northern Gulf of Mexico Hypoxic Zone." *Proceedings of the Royal Society B* 279(2011): 19–27. DOI: 10.1098/rspb.2011.0529

Thomas, P., M.S. Rahman, I.A. Khan, and J.A. Kummer. "Widespread Endocrine Disruption and Reproductive Impairment in an Estuarine Fish Population Exposed to Seasonal Hypoxia." *Proceedings of the Royal Society* B 274(2007): 2693-2701. DOI:10.1098/rspb.2007.0921.

Thronson, A. and A. Quigg. 2008. "Fifty-Five Years of Fish Kills in Coastal Texas." *Coastal and Estuarine Research Federation* 31(4/2008): 802-813. DOI 10.1007/s12237-008-9056-5.

Thurston, R. V. *Fish Physiology, Toxicology, and Water Quality Proceedings of the Sixth International Symposium.* La Paz, B.C.S. Mexico: Ecosystems Research Division. 2002.

Tian, D., W. Zheng, X. Wei, X. Sun, L. Liu, X. Chen, H. Zhang, Y. Zhou, H. Chen, H. Zhang, Q. Wang, R. Zhang, S. Jiang, Y. Zheng, G. Yang, and W. Qu. "Dissolved Microcystins in a Surface and Ground Waters in Regions with High Cancer Incidence in the Huai River Basin of China." *Chemosphere* 91(2013): 1064-1071. DOI:10.1016/j.chemosphere.2013.01.051.

Toledo-Lucas County Health Department. "Community Assessment for Public Health Emergency Response (CASPER) Following Detection of Microcystin Toxin in a Municipal Water Supply, Lucas County, Ohio." https://www.odh.ohio.gov/~/media/ODH/ASSETS/Files/eh/OHCASPERReport. Last accessed: January 8, 2016.

Tomer, M.D., S.A. Porter, K.M. Boomer, D.E. James, J.A. Kostel, M.J. Helmers, T.M. Isenhart, and E. McLellan. "Agricultural Conservation Planning Framework: Developing Multipractice Watershed Planning Scenarios and Assessing Nutrient Reduction Potential." *Journal of Environmental Quality* 44(2015): 754-767. DOI:10.2134/jeq2014.09.0386.

Trainer, V. L., L. Moore, B.D. Bill, N.G. Adams, N. Harrington, J. Borchert, D.A.M. da Silva, B.-T. L. Eberhart. "Diarrhetic Shellfish Toxins and Other Lipophilic Toxins of Human Health Concern in Washington State." Marine Drugs 11(2013): 1815-1835. DOI: 10.3390/md11061815.

Trainer, V.L., and F.J Hardy. "Integrative monitoring of marine and freshwater harmful algae in Washington State for public health protection." *Toxins* 7(2015): 1206-1234. DOI:10.3390/toxins7041206.

Turner, P.C., A.J. Gammie, K. Hollinrake, and G.A. Codd. "Pneumonia Associated with Contact with Cyanobacteria." *British Medical Journal* 300(1990): 1440-1441. DOI: 10.1136/bmj.300.6737.1440.

United States Department of Agriculture Natural Resources Conservation Service (USDA NRCS). *Assessment of the effects of conservation practices on cultivated cropland in the Great Lakes Region*. 2011. http://www.nrcs.usda.gov/Internet/FSE_DOCUMENTS/stelprdb1045480.pdf. Last accessed: May 15, 2015.

United States Environmental Protection Agency (EPA). *Hypoxia in the Northern Gulf of Mexico. An Update by the EPA Science Advisory Board.* Washington, DC: EPA Science Advisory Board. 2007. www.epa.gov/sab. Last accessed May 12, 2015.

United States Environmental Protection Agency (EPA). *Reactive Nitrogen in the United States: An Analysis of Inputs, Flows, Consequences, and Management Options*. Washington, DC: EPA Science Advisory Board.

2011. http://yosemite.epa.gov/sab/sabproduct.nsf/02ad90b136fc21ef85256eba00436459/67057225CC 780623852578F10059533D/$File/EPA-SAB-11-013-unsigned.pdf. Last accessed June 3, 2015.

United States Environmental Protection Agency (EPA). "National Coastal Condition Assessment (NCCA)." 2014. http://water.epa.gov/type/oceb/assessmonitor/ncca.cfm, Last accessed: April 29, 2015.

United States Environmental Protection Agency (EPA 2015a). *A Compilation of Cost Data Associated with the Impacts and Control of Nutrient Pollution.* EPA 820-F-15-096.2015. http://www2.epa.gov/sites/production/files/2015-04/documents/nutrient-economics-report2015.pdf. Last accessed: May 15, 2015.

United States Environmental Protection Agency (EPA 2015b). "Drinking Water Health Advisory for the Cyanobacterial Microcystin Toxins." EPA-820R15100. https://www.google.com/url?sa=t&rct=j&q=&esrc=s&source=web&cd=1&cad=rja&uact=8&ved=0ahUKEwibp_WhhKLKAhUFdj4KHdrHBhoQFggcMAA&url=http%3A%2F%2Fwww.epa.gov%2Fsites%2Fproduction%2Ffiles%2F2015-06%2Fdocuments%2Fmicrocystins-report2015.pdf&usg=AFQjCNE8s_eYfa1IhwJ-FyUnH-m525bK0A&bvm=bv.111396085,d.cWw. Last accessed: January 11, 2016.

United States Environmental Protection Agency (EPA 2015c). "Strategy Submitted to Congress to Meet the Requirements of P.L. 114-45." EPA-810R04003. http://www.epa.gov/sites/production/files/2015-11/documents/algal-risk-assessment-strategicplan-2015.pdf. Last accessed: January 13, 2016.

United States Geological Survey (USGS). 2002.

United States Geological Survey (USGS). "WaterQualityWatch – Continuous Real-Time Water Quality of Surface Water in the United States." 2014.http://waterwatch.usgs.gov/wqwatch/?pcode=00630. Last accessed: April 29, 2015.

United States Geological Survey (USGS). "National Water Information System: Web Interface, USGS 07374000 Mississippi River at Baton Rouge, LA." 2015. http://waterdata.usgs.gov/nwis/uv?site_no=07374000. Last accessed: April 29, 2015.

van der Schalie, W.H., H.S. Gardner, Jr, J.A. Bantle, C.T. De Rosa, R.A. Finch, J.S. Reif, R.H. Reuter, L.C. Backer, J. Burger, L.C. Folmar, and W.S. Stokes. "Animals as Sentinels of Human Health Hazards of Environmental Chemicals." *Environmental Health Perspectives*

107(1999): 309-315. http://www.ncbi.nlm.nih.gov/pmc/articles/PMC 1566523/pdf/envhper00509-0101.pdf.

Vanderploeg. H.A., J.R. Liebig, W.W. Carmichael, M.A. Agy, T.H. Johengen, G.L. Fahnenstiel, and T.F. Nalepa. "Zebra Mussel (*Dreisssena polymorpha*) Selective Filtration Promoted Toxic Microcystis Blooms in Saginaw Bay (Lake Huron) and Lake Erie." *Canadian Journal of Fisheries and Aquatic Sciences* 58(2001): 1208-1221. DOI: 10.1139/cjfas-58-6-1208.

Vaquer-Sunyer, R. and C.M. Duarte. "Thresholds of Hypoxia for Marine Biodiversity." *Proceedings of the National Academy of Sciences* 105(40/2008): 15452–15457DOI: 10.1073pnas.0803833105.

Vitousek, P.M., J.D. Aber, R.W. Howarth, G.E. Likens, P.A. Matson, D.W. Schindler, W.H. Schlesinger, and D.G. Tilman. "Human Alteration of the Global Nitrogen Cycle: Sources and Consequences." *Ecological Applications* 7(3/1997): 737-750. http://biol336- bowen.wikispaces.umb. edu/file/view/Vitousek+et+al.+1997.pdf.

Wells, M.L., V.L. Trainer, T.J. Smayda, B.S. O. Karlson, C.G. Trick, R.M. Kudela, A. Ishikawa, S. Bernard, A. Wulff, D.M. Anderson, W.P. Cochlan. In press. "Harmful Algal Blooms (HABs) and Climate Change: What do We Know and Where do We Go from Here?" *Harmful Algae* 49(2015). DOI:10.1016/j.hal.2015.07.009.

Westrick, J.A. "Cyanobacterial Toxin Removal in Drinking Water Treatment Processes and Recreational Waters." In *Cyanobacterial Harmful Algal Blooms: State of the Science and Research Needs*. New York: Springer. 2008. DOI: 10.1007/978-0-387-75865-7_13.

White, M.J., C. Santhi, N. Kannan, J.G. Arnold, D. Harmel, L. Norfleet, P. Allen, M. DiLuzio, X. Wang, J. Atwood, E. Haney, and M.V. Johnson. "Nutrient Delivery from the Mississippi River to the Gulf of Mexico and Effects of Cropland Conservation." *Soil and Water Conservation Society* 69(1/2014): 26-40. DOI: 10.2489/jswc.69.1.26.

Whitledge T.E. "Nationwide Review of Oxygen Depletion and Eutrophication in Estuarine and Coastal Waters. Executive Summary." *Report to U.S. Dept. of Commerce, NOAA*. Upton, New York: Oceanographic Sciences Division, Brookhaven National Laboratory. 1985. http://oceanservice. noaa.gov/websites/retiredsites/sotc_pdf/HYP.PDF.

Wilson, E. K. "Danger from Microcystins in Toledo Drinking Water Unclear." *Chemical Engineering News*, 92(32/2014). http://cen.acs.org/articles/ 92/i32/Danger-Microcystins-Toledo-WaterUnclear.html.

World Health Organization . *Guidelines for Drinking-Water Quality: Incorporating 1st and 2nd Addenda, Vol.1, Recommendations.* 3rd ed. Geneva, Swtizerland: WHO. 2008. http://www.who.int/water_sanitation_health/dwq/fulltext.pdf.

Wynne, T.T., R.P. Stumpf, M.C. Tomlinson, G.L.Fahnenstiel, J. Dyble, D.J. Schwab, S.J. Joshi. "Evolution of a Cyanobacterial Bloom Forecast System in Western Lake Erie: Development and Initial Evaluation." *Journal of Great Lakes Research* 39(2013): pp. 90-99. DOI: 10.1016/j.jglr.2012.10.003.

Yoder, J.S., B.G. Blackburn, G.F. Craun, V. Hill, D.A. Levy, N. Chen, S.H. Lee, R.L. Calderon, and M.J. Beach. "Surveillance for Waterborne-Disease Outbreaks Associated with Recreational Water – United States, 2001-2002." *Mobidity and Mortality Weekly Report Surveillance Summaries* 53(2004): 1-22. http://www.cdc.gov/mmwr/PDF/ss/ss5308.pdf.

Zhang, Y.-K. and K.E.Schilling. "Increasing Streamflow and Baseflow in Mississippi River since 1940s: Effect of Land Use Change." *Journal of Hydrology* 324 (2006): 412-422. DOI: 10.1016/j.jhydrol.2005.09.033.

ABBREVIATIONS

ALS	Amyotrophic Lateral Sclerosis (Lou Gehrig's Disease)
APEX	Agricultural Policy/Environmental eXtender
ARS	Agricultural Research Service (USDA)
ART	Analytical Response Team (NOAA)
ASP	Amnesic Shellfish Poisoning
BMAA	Beta-methylamine Alanine
BMP	Best management practice
CASPER	Community Assessment for Public Health Emergency Response
CCL	Contaminant Candidate List
CDC	Centers for Disease Control and Prevention
CEAP	Conservation Effects Assessment Project (USDA and others)
CFR	Code of Federal Regulations
CFP	Ciguatera Fish Poisoning
CHRP	Coastal Hypoxia Research Program
CRM	Certified Reference Materials
CWA	Clean Water Act

DHHS	Department of Health and Human Services
DOC	Department of Commerce
DOD	Department of Defense
DOI	Department of the Interior
DSP	Diarrhetic Shellfish Poisoning
ECOHAB	The Ecology and Oceanography of Harmful Algal Blooms program authorized by the Harmful Algal Bloom and Hypoxia Research and Control Act Centers for Disease Control and Prevention (1998, 2014)
EPA	United States Environmental Protection Agency
EQIP	Environmental Quality Incentives Program
FDA	Food and Drug Administration
GLRI	Great Lakes Restoration Initiative
GLOS	Great Lakes Observing System
HABs	Harmful Algal Blooms
HABHRCA	Harmful Algal Bloom and Hypoxia Research and Control Amendments Act of 2014
IOOS	Integrated Ocean Observing System
IWG-HABHRCA	Interagency Working Group on the Harmful Algal Bloom and Hypoxia Research and Control Act
LEOFS	Lake Erie Operational Forecast System
NASS	National Agricultural Statistics Service (USDA)
NASA	National Aeronautics and Space Administration
NCCOS	National Centers for Coastal Ocean Science (NOAA)
NCER	National Center for Environmental Research (EPA)
NGOMEX	Northern Gulf of Mexico Ecosystems and Hypoxia Assessment
NIEHS	National Institute of Environmental Health Sciences
NIFA	National Institute of Food and Agriculture (USDA)
NIST	National Institute of Standards and Technology
NOAA	National Oceanic and Atmospheric Administration
NORS	National Outbreak Reporting System
NPS	National Park Service
NRCS	Natural Resources Conservation Service (USDA)
NRL	Naval Research Lab
NSF	National Science Foundation
NSTC	National Science and Technology Council
OSTP	Office of Science and Technology Policy

PCM	Prevention, Control, and Mitigation
PCMHAB	Program that transitions promising PCM technologies and strategies to end-users and is authorized by the Harmful Algal Bloom and Hypoxia Research and Control Act (1998, 2014)
PMN	Phytoplankton Monitoring Network
PSP	Paralytic Shellfish Poisoning
RCPP	Regional Conservation Partnership Program
SPARROW	SPAtially-Referenced Regression on Watershed attributes model
SOST	Subcommittee on Ocean Science and Technology
SDWA	Safe Drinking Water Act
USACE	United States Army Corps of Engineers
USDA	United States Department of Agriculture
USGS	United States Geological Survey
WHO	World Health Organization

Glossary of Terms

Aerosolization – The process or act of converting some physical substance into the form of particles small and light enough to be carried on the air, i.e., into an aerosol.

Algae – Simple plant-like, photosynthetic organisms that form the base of most aquatic food webs, ranging from microscopic, single-celled diatoms, dinoflagellates, and cyanobacteria, to large seaweeds.

Aquatic – Of, in, or pertaining to marine and fresh waters, including the Great Lakes, and concentrates herein on those in the US and its territories.

Benthic – Ecological region at the lowest level of a body of marine or fresh water such as an ocean or a lake, including the sediment surface and sub-surface layers.

Best Management Practices – A method by which the adverse impacts of development and redevelopment are controlled through their application.

Carnivorous – Meat-eating animal.

Clean Water Act – This act establishes the basic structure for regulating discharges of pollutants into the waters of the United States and regulating quality standards for surface waters (1948, 1972).

Cyanobacteria – Photosynthetic bacteria that frequently form harmful algal blooms in marine and fresh waters; also called blue-green algae.

Cyanotoxin – Toxins produced by cyanobacteria.

Dissolved oxygen – The amount of gaseous oxygen present in the water.

Drinking Water Protection Act – An amendment to the Safe Drinking Water Act to provide for the assessment and management of the risk of algal toxins in drinking water, and for other purposes.

Ecosystem services – A benefit that an ecosystem provides to people.

Estuarine systems (Estuaries) – Systems that receive freshwater inputs from rivers, and that mix with ocean water.

Eutrophication – The enrichment of an ecosystem with chemical nutrients; typically compounds containing nitrogen, phosphorus, or both.

Farm Bill – The Agricultural Act of 2014; a five-year farm bill that reforms agricultural policy, reduces the deficit, and grows the economy.

Food web – Also known as a food chain. The visual depiction of relationships between living things, what they feed on, and what feeds on them. For example: Little fish eat algae. Bigger fish eat the little fish. Humans eat the bigger fish.

Freshwater – Naturally-occurring water on the Earth's surface in ice sheets, ice caps, glaciers, icebergs, bogs, ponds, lakes, rivers and streams, and underground as groundwater in aquifers and underground streams. Freshwater salinity is less than 0.5 g/kg.

Halocline – A vertical gradient of the salinity (saltiness) of a body of water.

Harmful algal blooms (HABs) – A small subset of algal species – including diatom, dinoflagellate, and cyanobacterial blooms – that produce toxins or grow excessively, harming humans, other animals, and the environment.

Herbivorous – Plant-eating animal.

Hypoxia – In reference to this report, waters that are, or have been, severely depleted of oxygen. **Hypoxic events** – When a body of water experiences a deficiency of oxygen.

In situ – In the normal location.

Nonpoint source – Pollution of a non-specific origin that is introduced into a body of water from land runoff, precipitation, atmospheric deposition, drainage, seepage, or hydrologic modification.

Nuisance algal blooms – Excessive algal/cyanobacterial growth that can promote pathogens and cause detrimental effects like hypoxia, but that do not produce toxins.

Plankton – Diverse assemblage of organisms that live in the water column and cannot swim against a current.

Phytoplankton – Minute plant-like organisms and other photosynthetic organisms – including cyanobacteria, diatoms, and dinoflagellates – that live in water and cannot swim against a current.

Phytoplankton bloom – Rapid increase or accumulation in the population of microscopic algae in a water body.

Point source pollution – Pollutant loads discharged at a specific location from pipes, outfalls, and conveyance channels from factories, confined animal feedlots, municipal wastewater treatment plants or individual waste treatment facilities.

Public Health Security and Bioterrorism Preparedness and Response Act – 2002 law that requires community drinking water systems serving populations of more than 3,300 persons to conduct assessments of their vulnerabilities to terrorist attack or other intentional acts, and to defend against adversarial actions that might substantially disrupt the ability of a system to provide a safe and reliable supply of drinking water. The EPA and water utilities are responsible for enhancing water sector security and developing response measures for potential threats to the nation's water supplies and systems.

Research Plan and Action Strategy – First report mandated by HABHRCA 2014. The Comprehensive HAB and Hypoxia Research Plan and Action Strategy will report on challenges related to HABs and hypoxia, ongoing and planned research, and the agencies' roles and responsibilities for evaluating and managing these issues.

Riparian Buffer Line – Trees, shrubs, or other vegetation lining the banks of a waterway that stabilize the ground, preventing erosion.

Safe Drinking Water Act – Originally passed in 1974, the main Federal law that ensures the quality of Americans' drinking water. Under SDWA, EPA sets standards for drinking water quality and oversees the states, localities, and water suppliers who implement those standards.

Salinity – Dissolved salt content of a body of water expressed as grams salt per kilogram of water.

Stratification – Segregation of the water column into distinct layers due to density differences that occur when the surface layer is warmer or has a lower salt content than the underlying water.

Thermocline – A vertical gradient of temperature. Water temperatures typically decline with depth, but often not uniformly. A thermocline is the depth at which small changes in depth lead to relatively large changes in temperature. The greater the temperature change across the thermocline, the more stable and distinct the upper layer of water.

Upwell/upwelling – The process by which warm, less-dense surface water is drawn away from along a shore by offshore currents and replaced by cold, denser water brought up from the subsurface.

End Notes

[1] Ecosystem services are the contributions that a biological community and its habitat provide to our day-to-day lives, including providing food sources, recreation opportunities, and carbon sinks.

[2] http://coastalscience.noaa.gov/research/habs/habhrca

[3] CRMs, for the purpose of this paper, are analysis frameworks for identifying types of toxins and their concentrations. CRMs also establish standards for the types and calibrations of instruments that are used for performing these analyses.

[4] http://ciler.snre.umich.edu/

[5] http://tidesandcurrents.noaa.gov/ofs/leofs/leofs.html

[6] http://www.glerl.noaa.gov/res/HABs_and_Hypoxia/lakeErieHABArchive/

[7] http://www.nrcs.usda.gov/wps/portal/nrcs/detail/national/technical/nra/ceap/ws/?cid=nrcs143_014160

[8] http://www.cdc.gov/nors/

In: Harmful Algal Blooms and Hypoxia
Editor: Roberta Alvarez
ISBN: 978-1-63485-399-6

Chapter 2

ALGAL BLOOM-ASSOCIATED DISEASE OUTBREAKS AMONG USERS OF FRESHWATER LAKES: UNITED STATES, 2009–2010*

Elizabeth D. Hilborn, DVM[1,†], Virginia A. Roberts, MSPH[2], Lorraine Backer, PhD[3], Erin DeConno[4], Jessica S. Egan, MPH[4], James B. Hyde[4], David C. Nicholas, MPH[4], Eric J. Wiegert[4], Laurie M. Billing, MPH[5], Mary DiOrio, MD[5], Marika C. Mohr, MS[5], F. Joan Hardy, PhD[6], Timothy J. Wade, PhD[1], Jonathan S. Yoder, MPH[2], and Michele C. Hlavsa, MPH[2]

[1]Environmental Public Health Div, Office of Research and Development, US Environmental Protection Agency

[2]Div of Foodborne, Waterborne, and Environmental Diseases, National Center for Emerging and Zoonotic Infectious Diseases, CDC

---

* This is an edited, reformatted and augmented version of an article that appeared in the Centers for Disease Control and Prevention's Morbidity and Mortality Weekly Report, Vol. 63, No. 1, January 10, 2014.

† Corresponding author: Elizabeth D. Hilborn, hilborn.e@epa.gov, 919-966-0658.

[3]Div of Environmental Hazards and Health Effects, National Center for Environmental Health, CDC
[4]New York State Dept of Health
[5]Ohio Dept of Health
[6]Washington State Dept of Health

Harmful algal blooms (HABs) are excessive accumulations of microscopic photosynthesizing aquatic organisms (phytoplankton) that produce biotoxins or otherwise adversely affect humans, animals, and ecosystems. HABs occur sporadically and often produce a visible algal scum on the water. This report summarizes human health data and water sampling results voluntarily reported to CDC's Waterborne Disease and Outbreak Surveillance System (WBDOSS) via the National Outbreak Reporting System (NORS) and the Harmful Algal Bloom-Related Illness Surveillance System (HABISS)[1] for the years 2009–2010. For 2009–2010, 11 waterborne disease outbreaks associated with algal blooms were reported; these HABs all occurred in freshwater lakes. The outbreaks occurred in three states and affected at least 61 persons. Health effects included dermatologic, gastrointestinal, respiratory, and neurologic signs and symptoms. These 11 HAB-associated outbreaks represented 46% of the 24 outbreaks associated with untreated recreational water reported for 2009–2010, and 79% of the 14 freshwater HAB–associated outbreaks that have been reported to CDC since 1978. Clinicians should be aware of the potential for HAB-associated illness among patients with a history of exposure to freshwater.

Eleven freshwater HAB–associated outbreaks that occurred in 2009 or 2010 were reported to CDC by New York, Ohio, and Washington. Freshwater HAB-associated outbreaks were defined as outbreaks where algal blooms were noted by state health or environmental investigators.[2] These outbreaks resulted in at least 61 illnesses, two (3%) hospitalizations, and no known deaths. Of 58 persons who reported seeking health care, seven (12%) reported a visit to an emergency department and 34 (59%) reported a visit to another type of health-care provider. Where demographic data were available, 34 ill persons were female (59%) and 38 (66%) were aged ≤19 years (Table 1). The median duration of an outbreak (time in days from the date of first reported exposure to the date of last reported exposure) was 9 days (range: 0–44 days).[3] Onset time (time in days from reported exposure to reported onset of health effects) was not available for each reported illness. Among six outbreaks, the earliest onset of signs and symptoms reportedly occurred within 1 day of exposure. Among five outbreaks, median onset was calculated based on data for 27 persons; median onset times ranged from half a day to 2 days. All 11 outbreaks were associated with recreational activities at freshwater lakes during June, July, or August.

Health effects varied among outbreaks (Table 2). Effects included dermatologic signs or symptoms such as rash, irritation, swelling, or sores (eight outbreaks); gastrointestinal signs or symptoms (eight); respiratory signs or symptoms (six); fever (five); headache (four); neurologic signs or symptoms (four); ear symptoms (five); and eye irritation (three).[4] Three routes of exposure to recreational water were reported: contact (nine outbreaks), ingestion (six), and inhalation of aerosols (four) (Table 2). In seven (78%) of the outbreaks for which contact exposure was reported, affected persons developed rash or skin irritation; in each of the outbreaks for which ingestion exposure was reported, affected persons had gastrointestinal signs or symptoms; and in three (75%) of the outbreaks for which inhalation exposure was reported, affected persons had respiratory signs or symptoms.

Cyanobacteria are common components of freshwater HABs and can produce cyanotoxins, which include potent hepatotoxins, neurotoxins, and dermatotoxins that can harm humans and animals [1]. Water testing practices varied among outbreak investigations (Table 3). Eight outbreak investigations included evaluation of cyanotoxins. Detections included microcystin (eight of eight investigations), anatoxin-a (three of four), saxitoxin (two of five), and cylindrospermopsin (two of five). Reported cyanotoxin concentrations varied over time. Four outbreak investigations included detection of multiple cyanotoxins; two of these investigations also revealed a potential association with bird, fish, or dog illness or deaths. Three outbreak investigations identified cyanobacteria and two identified potentially toxic cyanobacteria without quantification. One outbreak investigation included testing for bacterial indicators of fecal contamination. *Escherichia coli* was identified and quantified; samples exceeded U.S. Environmental Protection Agency (EPA) recreational water quality criteria of 126 CFU/100 mL.[5]

What is already known on this topic?

Cyanobacteria can form harmful algal blooms that might produce potent toxins in surface waters. Several studies have reported adverse human health effects associated with recreational water exposure to cyanotoxins and cyanobacteria blooms.

What is added by this report?

During 2009–2010 in the United States, 11 outbreaks associated with fresh water and harmful algal blooms affected at least 61 persons, resulting in two hospitalizations and no known deaths. Among 58 persons for whom data are available, seven (12%) visited an emergency department and 34 (59%) visited a health-care provider; 66% of affected persons overall were aged ≤19 years. This report suggests that the time to onset of effects might be rapid, that children might be at higher risk for illness, and that harmful algal bloom–associated outbreaks occur during the warmer months.

What are the implications for public health practice?

Untreated recreational waters with harmful algal blooms present a potentially severe health risk to humans and animals. Environmental control and prevention of blooms is needed. Efforts to identify and correctly ascertain algal bloom associated-illness will improve case detection and contribute to the development of evidence-based prevention strategies.

EDITORIAL NOTE

Outbreaks associated with freshwater HABs previously were reported infrequently; only three were reported to WBDOSS for 1978–2008.[6] Freshwater cyanobacteria blooms are most likely to form on warm, stable bodies of water that are rich in nutrients [1]. All outbreaks reported for 2009–2010 occurred in northern states during June–August. The World Health Organization (WHO) has proposed cell density guidelines for cyanobacteria bloom risk levels for recreational waters: 20,000 cells/mL is associated with risk for short-term adverse health outcomes; at 100,000 cells/mL, additional risk for long-term illness exists; and cyanobacterial scum formation in bathing areas is associated with the additional risk for "potentially severe health outcomes" [2].

This report highlights the challenges of recognizing HABassociated illness. During recreation in or alongside water with HABs, persons might experience multiple routes of exposure and multiple health effects. Reported exposure routes might reflect reported health effects rather than the true exposure route, given the difficulty of determining if exposure occurred via ingestion, inhalation, or contact. Previous reports of health effects associated with recreational water exposure to HABs include gastrointestinal, respiratory, eye, ear, and dermatologic effects; fever and neurologic effects are reported less frequently [3]. The nonspecific nature of these effects might make it difficult for health-care providers to identify HABassociated illness. Health-care providers should be aware that HAB-associated illness might present differently from other recreational water–associated illnesses and onset might occur soon after exposure. For example, four of the HAB-associated outbreaks were notable for associated neurologic symptoms or confusion/visual disturbance and HAB-associated illness might occur soon after exposure (Table 2). Among the 11 reported outbreaks, most persons reporting illness were aged ≤19 years. Children might be at higher risk for HAB-associated health effects because of more frequent exposure to and greater ingestion of recreational water [4].[7]

Animal deaths associated with HABs are sentinel events signaling potential risk for human illness associated with exposure to recreational waters. In two outbreaks, dead fish were reported by beach attendees and there were possibly associated dog deaths at each site. Biologic samples from affected persons and animals can be analyzed for cyanotoxins to improve exposure assessment; however, these assays are performed only by research or specialty laboratories at this time [5–7].

Table 1. Number and percentage of persons (n = 58*) affected by harmful algal bloom–associated waterborne disease outbreaks, by sex and age group — United States, 2009–2010

Characteristic	No.	(%)
Sex Female	34	(59)
Male	24	(41)
Age group (yrs)		
<1	0	—
1–4	4	(7)
5–9	16	(28)
10–19	18	(31)
20–49	12	(21)
50–74	8	(14)
≥75	0	—

* Three persons of unknown sex and age are not included.

Table 2. Reported exposure, health effects, and health-care use resulting from harmful algal bloom–associated waterborne disease outbreaks — United States, 2009–2010

Health-care use*			Reported health effects (no.)§
Outbreak (by state)	Cases Reported exposure† Health-care provider Emergency department Hospitalized	Gastrointestinal	General Dermatologic Eye/Ear Neurologic Respiratory
New York	2 1 0 0 Contact		Rash (2),
(Outbreak 1)			swelling (1),
			sores (1)
New York	2¶ 0 0 0 Contact		Watery Nasal
(Outbreak 2)			eyes (2) congestion (2)
New York	2 0 0 0 Contact		Rash (2)
(Outbreak 3)			
Ohio	3 1 0 0 Contact,	Abdominal	Dizziness (1), Rash (1), Neurologic Cough (1),
(Outbreak 4)	ingestion,	cramps (1),	headache (1), skin symptoms congestion (1),
	inhalation	diarrhea (1),	muscle irritation (1) (1) wheezing (1),
		anorexia (1)	aches (1), shortness of
			fatigue (1), breath (1)
			sore throat (2)
Ohio	19 19 0 0 Contact,	Vomiting (11),	Fever (2) Rash (6) Eye
(Outbreak 5)	ingestion,	nausea (11),	irritation
	inhalation	abdominal	(5),
		cramps (7),	earache (5)
		diarrhea (5)	
Ohio**	7†† 2 0 0 Contact,	Anorexia (2),	Fever (2), Rash (6), Visual Confusion Cough (1),
(Outbreak 6)	ingestion,	diarrhea (1),	fatigue (2), skin disturbance (1) wheezing (1)
	inhalation	nausea (1)	headache (1), irritation (1) (1),
			muscle/joint earache (1)
			pain (1),
			malaise (1),
			weakness (1),
			sore throat (1)
Ohio	9 3 2 0 Contact	Abdominal	Fever (2), Rash (8) Eye Neurologic Respiratory

Health-care use*			Reported health effects (no.)§
Outbreak (by state)	Cases Reported exposure† Health-care provider Emergency department Hospitalized	Gastrointestinal	General Dermatologic Eye/Ear Neurologic Respiratory
(Outbreak 7)		cramps (3),	headache (2) irritation (1), symptoms symptoms (1)
		diarrhea (3),	earache (1) (2),
		nausea (3),	tingling (2),
		vomiting (2)	confusion (1)
Ohio§§	8 5 5 1 Contact,	Nausea (5),	Fever (4), Skin Earache (2) Confusion Respiratory
(Outbreak 8)	ingestion,	vomiting (4),	headache (4), irritation (6), (3), symptoms (5),
	inhalation	diarrhea (4),	dizziness (1), rash (3) neurologic cough (2),
		abdominal	fatigue (3), symptoms wheezing (1),
		cramps (2),	malaise (1), (3) chest tightness
		anorexia (1)	back pain (1) (1)
Ohio	2 0 0 0 Contact,	Diarrhea (2),	
(Outbreak 9)	ingestion	vomiting (2)	
Washington	3 2 0 0 Unknown	Gastroenteritis (3)	Fever (1)
(Outbreak 10)			
Washington	4 1 0 1 Ingestion	Gastroenteritis (3)	Dermatologic Ear Respiratory
(Outbreak 11)			symptoms (1) symptoms symptoms (1)
			(1)

* Multiple levels of health care might have been accessed by a person (e.g., used emergency department and was hospitalized). No deaths were reported.

† Route(s) of exposure reported for the outbreak via the National Outbreak Reporting System.

§ Multiple health effects might be reported by a person. Information about each symptom might not have been available for all persons in an outbreak.

¶ Health-care information is unknown for two persons in this outbreak.

** Dog and fish illness or death also reported.

†† Health-care information is unknown for one person in this outbreak.

§§ Dog, fish, and bird illness or death also reported.

It is important to maintain surveillance for HAB-associated illnesses to understand their public health impact. In the United States, cyanobacteria and their toxins are unregulated drinking water contaminants [8]. No federal regulation of cyanobacteria nor cyanotoxins in drinking water or any recommended guideline values for recreational waters, no standardized methods for cyanotoxin detection in water or in biologic specimens, and no national monitoring programs for occurrence currently exist. However, many states rely on guidelines published by WHO [2]; others have derived their own risk assessments and developed guidelines to support public health decision-making, such as posting advisories or closing bodies of water to any use [3, 9]. A summary of guidelines from different countries for exposure to cyanobacteria and their toxins was recently published [10].These guidelines and reports include action levels that might be applied by local, regional, state, or tribal entities to reduce exposure of humans and animals to cyanobacteria and their toxins. Although HABISS has been discontinued, NORS will continue to provide a mechanism for national reporting of HAB-associated outbreaks.

Testing water samples for cyanotoxins can contribute to the investigation of HAB-associated illness. Microcystins were detected during all eight outbreak investigations in which cyanotoxin testing was performed. Microcystin concentrations of ≥20 μg/mL exceeded the WHO guideline for moderate health risks in four outbreaks (Table 3) [2]. During investigations of these outbreaks, saxitoxin, cylindrospermopsin, and anatoxin-a also were detected. To date, there are few reports of documented human exposure to these cyanotoxins in recreational waters and there are no United States or international public health guidelines on their concentrations. Notably, neurologic effects were reported in 75% of outbreaks that included detection of known neurotoxins (anatoxin-a and saxitoxin).

The findings in this report are subject to at least three limitations. First, outbreak detection varies among localities. Second, reporting is voluntary. Finally, the reports described here likely represent underreporting of freshwater HABassociated outbreaks.

This report represents a first attempt to summarize a group of freshwater HAB-associated recreational waterborne disease outbreaks. More resources are needed for improvements in risk characterization of cyanobacteria and cyanotoxins exposure, water monitoring for potentially toxic cyanobacteria, cyanotoxin analysis of water samples and biologic specimens, and case-finding for human illnesses associated with exposure to HABs in recreational waters. HAB-associated outbreaks will likely increase as warm eutrophic bodies of water become more common over time as predicted by development and climate projections. Better characterization of the occurrence of blooms, bloom-associated environmental conditions, and of human illness associated with exposure to algal blooms is needed to develop evidence-based prevention strategies (e.g., optimized control of nitrogen and phosphorus nutrient pollution). EPA, which regulates recreational water quality, and CDC can support local and state health jurisdictions to optimize national outbreak surveillance, and thus better inform prevention and control efforts by providing guidance on what epidemiological, clinical, and environmental data are needed to support detection and investigation.

ACKNOWLEDGMENTS

Mary Clifton, Carey Schwab, Ohio Bur of Infectious Disease; Kathleen Smith, Ohio Environmental Protection Agency. Kathryn MacDonald; Washington State Dept of Health.

REFERENCES

[1] World Health Organization. Toxic cyanobacteria in water: a guide to their public health consequences, monitoring and management. Ingrid Chorus I, Bartram J, eds. London, United Kingdom: E&FN Spon; 1999. Available at http://www.who.int/water_sanitation_health/resourcesquality/toxcyanbegin.pdf.

[2] World Health Organization. Guidelines for safe recreational water environments. Volume 1: coastal and fresh waters. Geneva, Switzerland: World Health Organization; 2003. Available at http://whqlibdoc.who.int/publications/2003/9241545801.pdf.

[3] Stone D, Bress W. Addressing public health risks for cyanobacteria in recreational freshwaters: the Oregon and Vermont framework. Integr Environ Assess Manag 2007;3:137–43.

[4] Dufour AP, Evans O, Behymer TD, Cantú R. Water ingestion during swimming activities in a pool: a pilot study. J Water Health 2006; 4:425–30.

[5] Furey A, Crowley J, Hamilton B, Lehane M, James KJ. Strategies to avoid the mis-identification of anatoxin-a using mass spectrometry in the

forensic investigation of acute neurotoxic poisoning. J Chromatogr A 2005;1:91–7.

[6] Gessner BD, Bell P, Doucette GJ, et al. Hypertension and identification of toxin in human urine and serum following a cluster of mussel-associated paralytic shellfish poisoning outbreaks. Toxicon 1997;5:711–22.

[7] Hilborn ED, Carmichael WW, Soares RM, et al. Serologic evaluation of human microcystin exposure. Environ Toxicol 2007;5:459–63.

[8] US Environmental Protection Agency. Water: contaminant candidate list 3 – CCL. Washington, DC: US Environmental Protection Agency; 2012. Available at http://water.epa.gov/scitech/drinkingwater/dws/ccl/ccl3.cfm.

[9] Butler N, Carlisle J, Linville R. Toxicological summary and suggested action levels to reduce potential adverse health effects of six cyanotoxins. Sacramento, CA: Office of Environmental Health Hazard Assessment, California Environmental Protection Agency; 2012. Available at http://www.swrcb.ca.gov/water_issues/programs/peer_review/docs/calif_cyanotoxins/draft_cyanotoxinseport_oeha.pdf.

[10] Chorus I (ed.). Current approaches to cyanotoxin risk assessment, risk management and regulations in different countries. Dessau-Rosslau, Germany: Federal Environment Agency (Umweltbundesamt); 2012. Available at http://www.umweltdaten.de/publikationen/fpdf-l/4390.pdf.

End Notes

[1] The following states received grants to participate in HABISS: Florida, Iowa, Maryland, Massachusetts, New York, Oregon, South Carolina, Virginia, Washington, and Wisconsin.

[2] Two criteria must be met for a health event to be defined as an outbreak associated with recreational water: two or more persons must be linked epidemiologically, and the epidemiologic evidence must implicate recreational water as the probable source of illness. The reporting form and guidance for reporting via NORS are available at http://www.cdc.gov/nors. Outbreaks can be voluntarily reported to CDC by public health agencies in U.S. states, the District of Columbia, Guam, Puerto Rico, the Republic of the Marshall Islands, the Federated States of Micronesia, the Commonwealth of the Northern Mariana Islands, the Republic of Palau, and the U.S. Virgin Islands.

[3] Based on information for 10 outbreaks; data missing for one outbreak.

[4] More than one type of health effect or exposure could be reported by a single person

[5] Additional information available at http://water.epa.gov/scitech/swguidance/standards/criteria/health/recreation/upload/factsheet2012.pdf.

[6] One outbreak in New Hampshire (August 2001) in which Oscillatoria was suspected and isolated from a lake in a state park, and gastrointestinal symptoms were reported in 42 children; and two outbreaks in Nebraska (July 2004) involving microcystin from a lake, in which respiratory and dermatologic symptoms were reported in 22 persons. Additional information available at http://www.cdc.gov/healthywater/surveillance/rec-watersurveillance-reports.html.

[7] Data available at http://www.census.gov/compendia/statab/2012/tables/ 12s1249.pdf.

INDEX

A

B

C

D

H

I

L

M

N

O

P

Q

R

S

T

U

V

W

Z